# The Cognitive Approach in Cloud Computing and Internet of Things Technologies for Surveillance Tracking Systems

Intelligent Data Centric Systems: Sensor Collected Intelligence

# The Cognitive Approach in Cloud Computing and Internet of Things Technologies for Surveillance Tracking Systems

Edited by

**Dinesh Peter**

*Department of Computer Science and Engineering, Karunya Institute of Technology and Sciences, Coimbatore, India*

**Amir H. Alavi**

*Department of Civil and Environmental Engineering, The University of Pittsburgh, PA, USA*

**Bahman Javadi**

*School of Computing, Engineering and Mathematics, Western Sydney University, Sydney, NSW, Australia*

**Steven L. Fernandes**

*Department of Computer Science, The University of Central Florida, Orlando, FL, USA*

Series Editor

**Fatos Xhafa**

Academic Press is an imprint of Elsevier
125 London Wall, London EC2Y 5AS, United Kingdom
525 B Street, Suite 1650, San Diego, CA 92101, United States
50 Hampshire Street, 5th Floor, Cambridge, MA 02139, United States
The Boulevard, Langford Lane, Kidlington, Oxford OX5 1GB, United Kingdom



**British Library Cataloguing-in-Publication Data**
A catalogue record for this book is available from the British Library

**Library of Congress Cataloging-in-Publication Data**
A catalog record for this book is available from the Library of Congress

ISBN: 978-0-12-816385-6

For Information on all Academic Press publications
visit our website at https://www.elsevier.com/books-and-journals

*Publisher:* Mara Conner
*Acquisitions Editor:* Chris Katsaropoulos
*Editorial Project Manager:* Ali Afzal-Khan
*Production Project Manager:* Maria Bernard
*Cover Designer:* Christian Bilbow

Typeset by MPS Limited, Chennai, India

# Contents

# List of Contributors

**S. Chandrakala**
Department of Computer Science and Engineering, School of Computing, SASTRA Deemed to be University, Thanjavur, India

**Renu Mary Daniel**
Department of Computer Science and Engineering, Karunya Institute of Technology and Sciences, Coimbatore, India

**K. Deepak**
Department of Computer Science, School of Computing, SASTRA Deemed to be University, Thanjavur, India

**P. Deepan**
Department of Computer Science and Engineering, Annamalai University, Chidambaram, India

**Sharmila Anand John Francis**
Department of Computer Science, King Khalid University, Abha, Saudi Arabia

**Deva Priya Isravel**
Department of Computer Science and Engineering, Karunya Institute of Technology and Sciences, Coimbatore, India

**Titus Issac**
Karunya Institute of Technology and Sciences, Coimbatore, India

**T. Anita Jones**
Department of Electronics and Communication Engineering, Karunya Institute of Technology and Sciences, Coimbatore, India

**Aldrin Karunakaran**
Department of Process Engineering, International Maritime College, Sultanate of Oman

**P.U. Krishnanugrah**
Federal Institute of Science and Technology, Angamaly, India

**G. Pradeep Kumar**
Velammal College of Engineering and Technology, Madurai, India

**R.J.S. Jeba Kumar**
Department of Electronics and Communication Engineering, Karunya Institute of Technology and Sciences, Coimbatore, India

**Murugan Mahalingam**
Department of Electronics and Communication Engineering, Valliammai Engineering College, Chennai, India

**K. Mahesh**
Department of Computer Applications, Alagappa University, Karaikudi, India

**S. Malini**
Department of Computer Science and Engineering, School of Computing, SASTRA Deemed to be University, Thanjavur, India

**Anitha Mary**
Department of Instrumentation Engineering, Karunya Institute of Technology and Sciences, Coimbatore, India

**D. Mohanapriya**
Department of Computer Applications, Alagappa University, Karaikudi, India

**Vijay Rajeev**
Federal Institute of Science and Technology, Angamaly, India

**Revathi Arumugam Rajendran**
Department of Information Technology, Valliammai Engineering College, Chennai, India

**Elijah Blessing Rajsingh**
Department of Computer Science and Engineering, Karunya Institute of Technology and Sciences, Coimbatore, India

**Preethi Sambandam Raju**
Department of Electronics and Communication Engineering, Valliammai Engineering College, Chennai, India; Department of Information and Communication Engineering, Anna University, Chennai, India

**Lina Rose**
Department of Instrumentation Engineering, Karunya Institute of Technology and Sciences, Coimbatore, India

**S. Roshan**
Department of Computer Science, School of Computing, SASTRA Deemed to be University, Thanjavur, India

**K.S. Senthilkumar**
Department of Computers and Technology, St. George's University, Grenada, West Indies

**N. Shreyas**
Department of Computer Science and Engineering, School of Computing, SASTRA Deemed to be University, Thanjavur, India

**Salaja Silas**
Department of Computer Science and Engineering, Karunya Institute of Technology and Sciences, Coimbatore, India

**P. Sreevidya**
Federal Institute of Science and Technology, Angamaly, India

**B. Sridevi**
Velammal Institute of Technology, Chennai, India

**G. Srivathsan**
Department of Computer Science, School of Computing, SASTRA Deemed to be University, Thanjavur, India

**L.R. Sudha**
Department of Computer Science and Engineering, Annamalai University, Chidambaram, India

**D. Sugumar**
Department of Electronics and Communication Engineering, Karunya Institute of Technology and Sciences, Coimbatore, India

**G. Thennarasi**
Department of Electronics and Communication Engineering, Karunya Institute of Technology and Sciences, Coimbatore, India

**S. Veni**
Amrita Institute of Science and Technology, Coimbatore, India

**M. Venkatraman**
Department of Computer Science and Engineering, School of Computing, SASTRA Deemed to be University, Thanjavur, India

CHAPTER

# RELIABLE SURVEILLANCE TRACKING SYSTEM BASED ON SOFTWARE DEFINED INTERNET OF THINGS

1

**Deva Priya Isravel, Salaja Silas and Elijah Blessing Rajsingh**

*Department of Computer Science and Engineering, Karunya Institute of Technology and Sciences, Coimbatore, India*

## CHAPTER OUTLINE



## 1.1 INTRODUCTION

Intelligent surveillance tracking system provides real-time and sustained monitoring of a person, groups of people, objects, behavior, events, or environment. In recent years, there has been a rise in the use of surveillance tracking system for numerous applications. They are widely used in military applications, public monitoring, and commercial purposes. The main purpose of these kinds of observation is to provide personal and public safety, identify crime, prevent criminal activity, and enhance businesses and scientific research. Surveillance is extensively used for monitoring the safety of people from street corners to crowded places such as railways, airports, restaurants, malls, etc. It is also widely used in health-care services for observing patients and hospital facilities to provide quality care and support emergency preparedness and emergency services [1,2]. A lot of businesses use surveillance to boost their company productivity and profit by monitoring less their employees and concentrating more on businesses.

The Cognitive Approach in Cloud Computing and Internet of Things Technologies for Surveillance Tracking Systems.
DOI: https://doi.org/10.1016/B978-0-12-816385-6.00001-5

The benefits of a surveillance system can be applied in a number of ways other than security purpose. They are applied for building smart home automation and smart city projects [3–5]. The recent interest in mass surveillance for various causes introduces increased complexity in managing the surveillance system. With the recent advancement of new disruptive technologies, reliable and sophisticated surveillance tracking is built with multiple features. The surveillance tracking system should be able to provide a fast, time-sensitive, reliable, and rapid recovery mechanism for monitoring and predicting possible dangerous situations. The scale and complexity of surveillance networks are approaching massive and rapid changes. With the rise in the proportion of Internet of Things (IoT) enabled devices, sensors, mobile devices, smartphones, etc., the total Internet traffic has grown tremendously. A number of devices communicating at the same time with the base station increase and congestion occurs in the surveillance network. The amount of traffic exchanged across devices is also huge. Managing huge volumes of data traffic generated from multiple monitoring and capturing devices is complex, because it has to be processed simultaneously and sent to the appropriate base station or to the cloud for further investigation and data analytics. Because of this, the traditional network architecture has huge complexity and challenges in handling the network traffic and network management. Therefore improvising the intelligent and automated surveillance tracking system requires the scientific and research community to provide solutions. To address the challenges faced by traffic management, a new software defined networking (SDN) technology can be integrated into the surveillance tracking system to enhance the data transmission concerns that exist in the legacy surveillance network.

Section 1.2 presents in detail the concepts of the surveillance tracking system. Section 1.3 discusses the various communication technologies that are already in use to deploy the surveillance system. Section 1.4 provides a brief overview of the SDN technologies and its benefits when combined with the IoT. The novel framework of SDN-assisted IoT solution for building an effective reliable surveillance system is discussed in Section 1.5. Finally, Section 1.6 provides the conclusion.

## 1.2 SURVEILLANCE TRACKING SYSTEM

The surveillance tracking system is a system that is used for tracking humans, objects, vehicles, etc. and monitoring environment for ensuring safety and avoiding intruders. The surveillance has become a necessity for monitoring public and private spaces. Modern surveillance systems have demanding requirements with enormous, busy, and complex scenes, with heterogeneous sensor networks. The real-time acquisition and interpretation of the environment and flagging potentially critical situations are challenging [6]. The implementation of the surveillance system has three major phases. They are data capturing, data analysis, and postprocessing. In the data capturing phase, the web traffic, audio/video, and VoIP contents are captured from the environment and given as input to the preprocessing module for extraction. The data analysis phase comprises different steps in processing to obtain an enhanced quality image. The steps are image preprocessing, object-based analysis, event-based analysis, and visualization. In image preprocessing, video frames are extracted from the captured visual. Then interframes are estimated and image encoding is applied. The subregions of the image are identified by segmenting the image into partitions of different configurations in order to detect the person. The second phase of object-based analysis involves person tracking, posture classification, and body-based analysis. Then, estimations are then updated. The event-based analysis contains interaction modeling and activity analysis to explore the events happening. Finally, in the visualization stage, based on the camera calibration, an enhanced quality image is obtained. After the analysis phase, the extracted image is sent for postprocessing to take further evaluations and generated actions. Fig. 1.1 depicts the three major phase of the surveillance system.

### 1.2.1 CLASSIFICATION OF THE SURVEILLANCE

The surveillance tracking system can be broadly classified into three types. They are audio surveillance, video surveillance, and Internet surveillance.

#### *1.2.1.1 Audio surveillance*

Audio surveillance involves listening to sounds and detecting various acoustic events. Audio surveillance is applied to a wide range of applications like spying, patrolling, detective operations, etc. A number of sophisticated devices are available to work under different circumstances. Some of the listening devices are telephones, microphones, smartphones, wiretapping, voice recorders, and acoustic sensors. These devices capture the sound and then are analyzed to detect unusual and unsafe events [7]. The two important things in audio surveillance are feature extraction and audio pattern recognition [8].

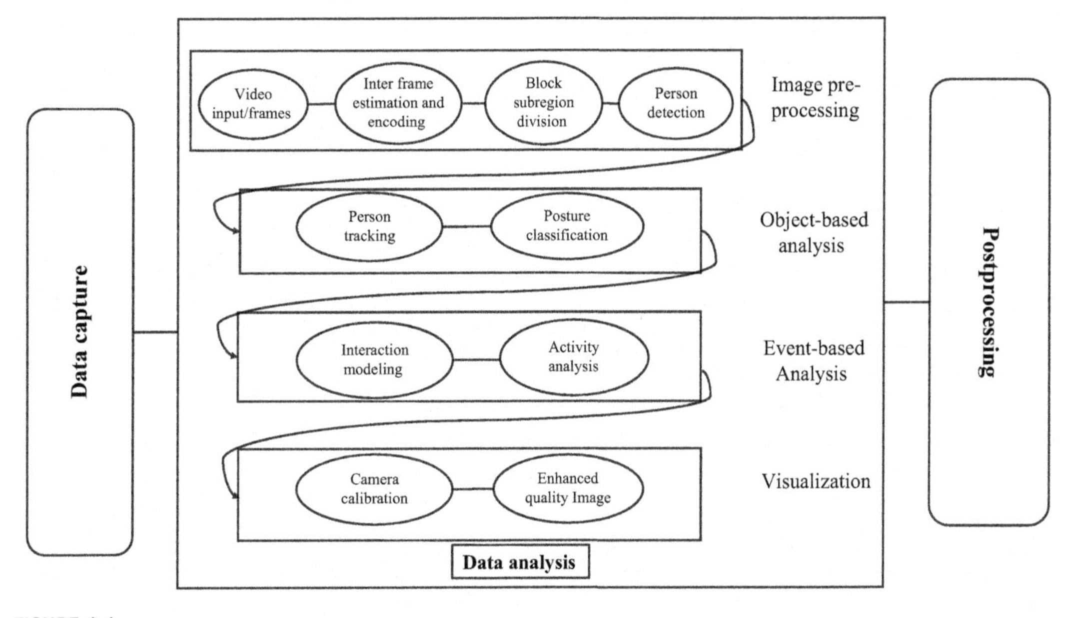


**FIGURE 1.1**

Phases of surveillance tracking system.

#### *1.2.1.2 Video surveillance*

Video surveillance is monitoring the behavior or activity in an area by capturing video images, and these images are transferred to the automated system for further processing. The devices used are cameras, sensors, high definition capability video capturing devices, and display monitors to view the captured video in real time. Earlier days, the video surveillance system used simple video acquisition and display systems. But with the advancement in technologies, modern video surveillance tracking system has sophisticated devices for image and video acquisition and data processing. It can integrate image and video analysis algorithms for pattern recognition, decision-making, and image enhancement. The major task in video surveillance is the detection and recognition of moving objects, tracking, performing behavioral analysis, and retrieving of the important data of concern [9]. Extracting visual from long footages is a laborious and time-consuming task. Therefore visual analytics is required to process visual content without human intervention. Various tools and technologies are integrated to understand the different dimensions of video summarization, visualization, interaction, and navigation [10]. A lot of challenges exist in controlling and monitoring the visual while streaming live from a large number of video surveillance cameras [11].

#### *1.2.1.3 Internet surveillance*

Internet surveillance is monitoring online and offline computer activity. It involves monitoring the exchange of digital data across the Internet. Here, monitoring is often carried out covertly by government agencies, service providers, and cybercriminals. As the Internet has become part of everyday life, surveillance helps to identify, disrupt, and mitigate the misuse of the Internet by attackers and criminals. Internet surveillance by multiple intelligent services has become a powerful tool in monitoring individuals globally. But at the same time, the privacy and convenience of the users are intruded [12]. There is a huge challenge in handling massive data volumes in terms of collection, storage, and analysis.

### 1.2.2 APPLICATIONS

There are numerous environments and areas where the surveillance tracking system can be applied to meet various needs. But some of the most commonly applied areas of surveillance are discussed below. Fig. 1.2 enumerates each surveillance functionality.

#### *1.2.2.1 Corporate surveillance*

The corporate surveillance is monitoring the happenings in public places like shopping malls, bus stands, railway stations, airports, restaurants, and people gathering using closed-circuit television (CCTV) cameras to ensure the safety of the people. Here, the devices are positioned at a fixed location to monitor only a particular area of interest [13]. In the earlier system, the data collected continuously from these CCTV cameras were observed by a human for any theft, or any untoward incident. But with sophisticated technologies, automated intelligent algorithms are used for identification, recognition, and image analysis. Extracting useful information from the long hours of video footage is a tedious task. Also transferring the data to a data center or cloud for storage and

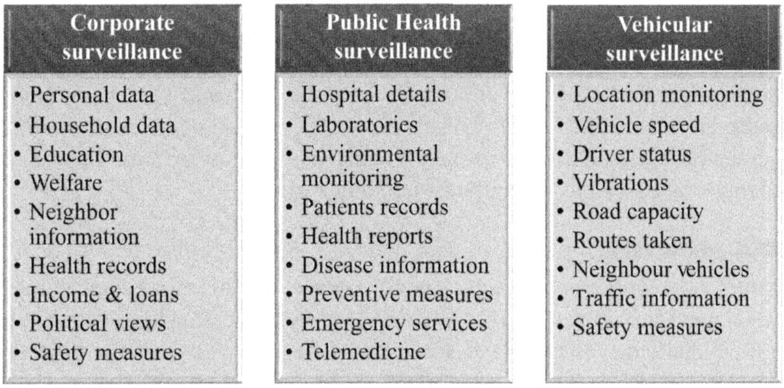

**Corporate surveillance**

- Personal data
- Household data
- Education
- Welfare
- Neighbor information
- Health records
- Income & loans
- Political views
- Safety measures

**Public Health surveillance**

- Hospital details
- Laboratories
- Environmental monitoring
- Patients records
- Health reports
- Disease information
- Preventive measures
- Emergency services
- Telemedicine

**Vehicular surveillance**

- Location monitoring
- Vehicle speed
- Driver status
- Vibrations
- Road capacity
- Routes taken
- Neighbour vehicles
- Traffic information
- Safety measures

**FIGURE 1.2**

Surveillance functionality.

retrieval for further processing is challenging and requires reliable connectivity between all the devices. Delay in request or response may lead to fatal situations.

#### *1.2.2.2 Public health surveillance*

Public health deals with systematic health-related data collection, analysis, interpretation, and dissemination for providing quality health care of the public. This is also called syndrome surveillance. It helps in diagnosing the disease and supporting real-time assistance. Timely dissemination of data helps to prevent and control the disease from further spreading [14]. The surveillance network should provide accurate information even from distant geographical locations to prevent the outbreak of disease. It also involves the observation of patients for symptoms and vital signs, interprets clinical changes, and notifies the response team to offer treatment to patients.

#### *1.2.2.3 Vehicular surveillance*

Vehicular surveillance involves monitoring and tracking of vehicle movements. Not only that, it has to monitor the pedestrians on the road. The task is to have control of the transportation network and ensure safety and hassle-free driving [15]. Now with autonomous car technology, the surveillance tracking system should be effective and intelligent in order to ensure no accidents. Continuous monitoring of the environment is required to avoid congestion and take the optimal path for reaching the destination [16]. Surveillance tracking systems collect information about the vehicle location, amount of fuel quantity, tire pressure, engine temperature, vehicle speed, driver's activities, etc. The collected data are sent to the server via mobile or satellite for traffic analysis and evaluation purposes [17]. This surveillance when implemented effectively will help to avoid accidents and pollution, and provide a fast and safe journey.

In a rapidly evolving surveillance tracking system, several challenges arise with respect identifying, tracking, and processing. Devising complex models with an intricate set of equipment makes the tracking system complex and challenging. The most common challenges are discussed in Section 1.2.3.

### 1.2.3 CHALLENGES

There has been an unprecedented level of growth in the usage of a surveillance tracking system for a good cause, such as public health surveillance, disease surveillance, vehicle surveillance, animal surveillance, marketing monitoring surveillance, and so on. But along with it, there are few challenges that are essential to be addressed in the surveillance tracking system. This section briefs on the various challenges faced by the surveillance tracking system.

#### 1.2.3.1 Dynamic processing

This is one of the most important factors when it comes to the surveillance tracking system. Despite the advancement in various Internet technologies, dynamic real-time delivery of data to the responders is still a challenge. Even if the surveillance devices collect from different sources in real-time but if it cannot deliver the processed data on time then the surveillance will not be effective for handling emergency and critical situations.

#### 1.2.3.2 Visual processing

Video processing is essential in order to understand the activity going on in the environment. To gain knowledge on the visual, the foreground visual and the background visual have to be extracted [18]. New advent methodologies are required to get trajectory models, categorization, recognition, behavior analysis, etc. The validity and accuracy of a surveillance system can be improved by adding video analytics and cloud infrastructure. But all this introduces new challenges in terms of cost and response time.

#### 1.2.3.3 Data management

As the surveillance system grows with a collection of heterogeneous devices, managing the data collected continuously from all the devices is challenging. Extracting, transmission, and retrieval of important data from the huge volume of raw data are difficult and time-consuming. If the data collected are not properly analyzed, then deep insight into the happening cannot be understood and intelligent reasoning may not be available to offer effective solutions for emergency and critical situations.

#### 1.2.3.4 Security and privacy

In most of the surveillance environments, the end user is unaware of being monitored. When individuals offer their personal information, it can be misused or used for threatening. Individual freedom is interrupted in some cases if the surveillance data collected end up with the wrong person hands and then the security and privacy of the individuals are lost. Though different surveillance system has a different regulation mechanism for ensuring the safety of the public, there is always caution on the usages of personal and private information.

## 1.3 WIRELESS COMMUNICATION TECHNOLOGIES

A number of technologies are utilized to facilitate the communication of captured data in the surveillance tracking system. Most often, the wireless low-cost portable technologies are used because of its convenience, cost, and easy availability. The various wireless technologies that enable surveillance are radiofrequency identification (RFID), near field communication (NFC), global positioning system (GPS), Bluetooth, ZigBee, and Wi-Fi. Fig. 1.3 shows the wireless technologies that are most commonly used.

- *RFID*: It is used to identify and track tags attached to the objects. These tags communicate using electromagnetic fields and need not be in the line of sight of the reader. These tags are widely used in industries for tracking objects and assets. They can be attached to automobiles, pet animals, pharmaceuticals, clothing, etc. to locate them easily.
- *NFC*: It is a short-range wireless technology that is used in electronic devices such as a smartphone, sensors, credit cards, and so on. This NFC enabled devices help to identify documents and objects within their communication range.
- *Bluetooth*: It is a technology that uses radiofrequency for transmitting data between Bluetooth enabled devices. It is low-power high-speed wireless technology that is used in phones, computers, networking devices, printer, mouse, etc. This technology is also used in a surveillance system for short-range transmission. It is widely used in the indoor environment for tracking customers, studying the movement of objects and individuals [19].
- *ZigBee*: It is a standard that is used in personal area network for home automation. It is mainly designed to enable low-cost, low-power standard for monitoring and controlling objects with a

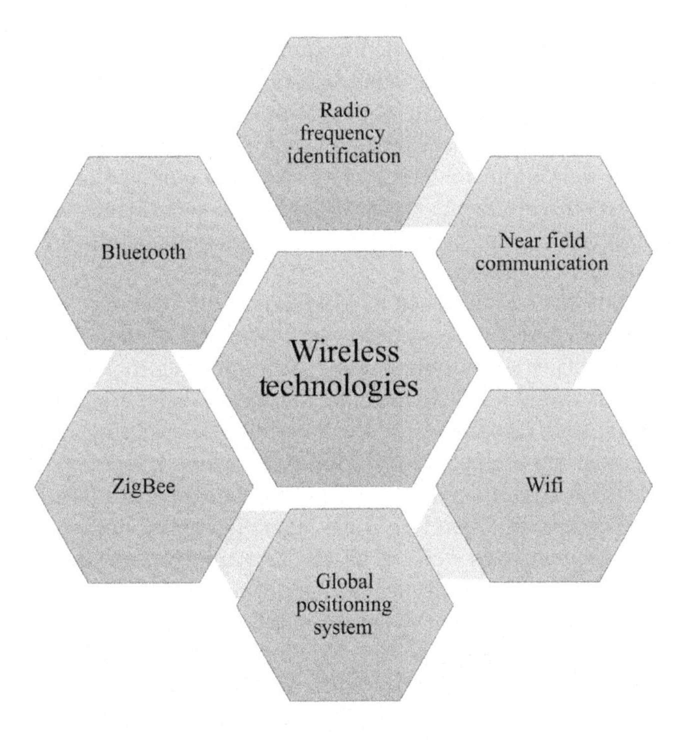


**FIGURE 1.3**

Wireless communication technology.

small geographic location. This technology is used along with sensors and IoT devices for transmission of data.

- *GPS*: It uses satellite communication to find the ground position of the object. The GPS receivers are installed in smartphones, automobiles, and other handheld devices. It is used as a navigation system by the civilians. It is used widely for vehicular tracking and to provide navigation and driving instructions.
- *Wi-Fi*: It is a wireless technology that uses radio waves and provides high-speed connectivity. It enables devices to connect to the Internet seamlessly. The surveillance tracking system uses Wi-Fi technology for transferring the captured data from the device to the display monitors, processing module, and then even stored in the cloud for long-term analysis. Internet surveillance can be carried out by probing Wi-Fi traces in any infrastructure to collect the individual's data, movements, environment, resource utilization, etc.

Though these wireless technologies play a vital role to establish communication across small and large networks, still technical difficulty exists in fulfilling the accessibility requirements and the recurring challenges of QoS, reliability, and medium access.

## 1.4 SOFTWARE DEFINED NETWORKING

As most of the time the wireless communication medium is utilized for transferring data, the reliability is not effectively ensured. When the huge volume of video footage is streamed across the network for taking a constructive decision, it is mandatory for the network to support fast transmission and quick response to actions. The existing system of operation is very rigid and vendor-specific. Customizing the working of devices is very hard and complex as it requires proprietary middleware software to reconfigure them. The next-generation surveillance tracking system can be built with a new network architecture by using SDN to improve parameters like network congestion, throughput, latency, and QoS parameters.

SDN is a new paradigm that was designed to provide network operators with more control over how traffic is forwarded from one place to another place. The idea of the SDN is to provide a separation between the control plane and the forwarding plane. It simplifies network management by providing global network visibility. SDN architecture has three layers. They are control plane layer, data plane layer, and application layer. Fig. 1.4 illustrates the responsibility of each layer.

The centralized controller has a global view of the entire network topology. The controller decides based on the request that comes from the forwarding plane devices. The network intelligence resides in the centralized controller. Therefore if any reconfigurations are required, then it is sufficient to configure only the controller and not all the forwarding plane devices. The forwarding plane receives instructions from the controller and performs only forwarding of the traffic. The administrator is able to customize the policies and protocols across the network devices via the centralized controller. The SDN supports scalability without affecting the reliability and performance by offering network programmability. The resource utilization and energy management can be improved by dynamic autoreconfiguration of the devices from the controller. The SDN offers traffic engineering capabilities for different types of traffic and enables congestion-free reliable transmission. The key benefits of integrating SDN with IoT (SDIoT) are shown in Table 1.1.

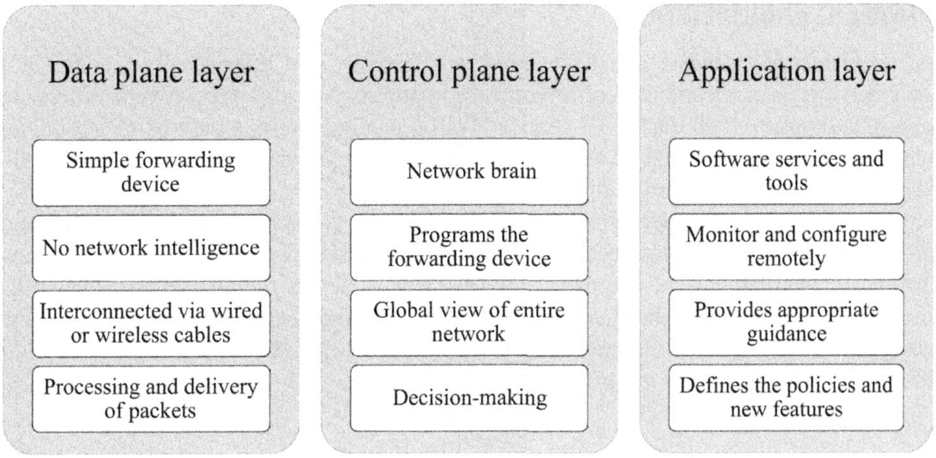

| Data plane layer | Control plane layer | Application layer |
|---|---|---|
| Simple forwarding device | Network brain | Software services and tools |
| No network intelligence | Programs the forwarding device | Monitor and configure remotely |
| Interconnected via wired or wireless cables | Global view of entire network | Provides appropriate guidance |
| Processing and delivery of packets | Decision-making | Defines the policies and new features |

**FIGURE 1.4**

Software defined networking layers.

**Table 1.1 Benefits of software defined networking with Internet of Things (SDIoT).**

| Key Benefits of SDIoT | Description |
|---|---|
| Global visibility | The centralized controller has a global database and has knowledge of the entire network topology. It knows the current status and behavior of the network |
| Programmability | A well-defined API is available to program these applications based on the requirements. The forwarding plane devices can be programmed to get the desired network behavior |
| Dynamic management | The traffic flow can be steered dynamically by the controller to offer flexibility based on decision policies and rules |
| Reduce complexity | Simplifies network management by enabling autoreconfigurations, agility, and flexibility |

## 1.5 SOFTWARE DEFINED SURVEILLANCE TRACKING SYSTEM

Today's surveillance tracking system requires the underlying network architecture to behave in real time and to scale up to a large amount of traffic. The network architecture should be capable of classifying a variety of traffic types for different applications, and to provide an appropriate and specific service for each traffic type in a very short time period such as within milliseconds. Highly efficient network management is desirable to significantly improve resource utilization for optimal system performance when it encounters the rapid growth and demand for massive-scale processing with audio and video data. To augment these features, a novel SDN-based traffic engineering framework for SDIoT is proposed.

### 1.5.1 TRAFFIC ENGINEERING

SDN traffic engineering scheme can be utilized to adaptively and dynamically manage or route traffic in a network to accommodate different traffic patterns and improve network efficiency and performance, and ensure high quality of service. Traffic engineering is a method of optimizing the performance of a telecommunication network by dynamically analyzing, predicting, and regulating the behavior of data transmitted over that network.

A major problem with the underlying surveillance network is the dynamic nature of the network applications running on all the connected devices and their environments [20,21]. This means that the performance requirements of the data flows exchanged across the surveillance network vary over time. Therefore to ensure the quality of service (QoS) and real-time delivery of data, traffic engineering schemes have to be incorporated with the SDIoT [22]. The various objectives of applying traffic engineering principles are given in Fig. 1.5.

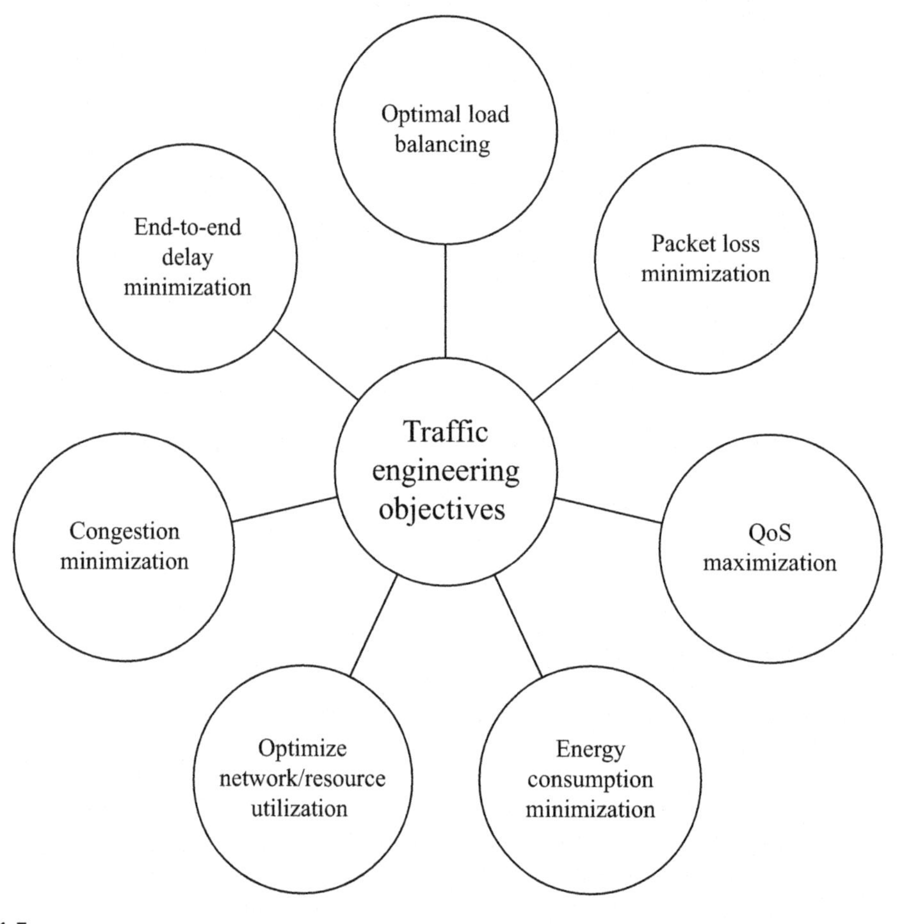


**FIGURE 1.5**

Traffic engineering objectives.

### 1.5.2 PROPOSED TRAFFIC ENGINEERING FRAMEWORK

The challenges that are especially faced in video surveillance can be overcome to a considerable extent by implementing the traffic engineering mechanism. The working of the surveillance network with the SDN integrated is depicted as a block diagram in Fig. 1.6. The devices that form the surveillance systems are connected by means of a wired or wireless medium. Initially, the devices are configured with default settings and policies based on the requirements. There is a centralized controller representing the control plane and has the knowledge of the devices that are part of the surveillance network. The captured data traffic can be in the form of web, audio, or video traffic. The current traffic flow statistics and network status are communicated to the controller periodically. This keeps the controller updated on the network status. The traffic engineering principles are applied by the controller. Based on the various application demands that arrive from the surveillance tracking system, the devices are auto reconfigured in real time by the controller to accommodate new demands. The reconfiguration of devices involves modification to the flow table entries in data plane devices. On applying the traffic engineering mechanism, the forwarding device flow table entries are dynamically updated. If there are no changes to the flow table, then the existing policies are applied for routing the traffic. The continuous monitoring by the SDN enabled controller helps to dynamically find optimal path for traffic.

In the earlier days, the captured audio visual and video visual of the events were recorded. To analyze and understand, the recorded video was replayed and manually checked by personnel. But in modern surveillance tracking system, the visuals are continuously and systematically sent to the

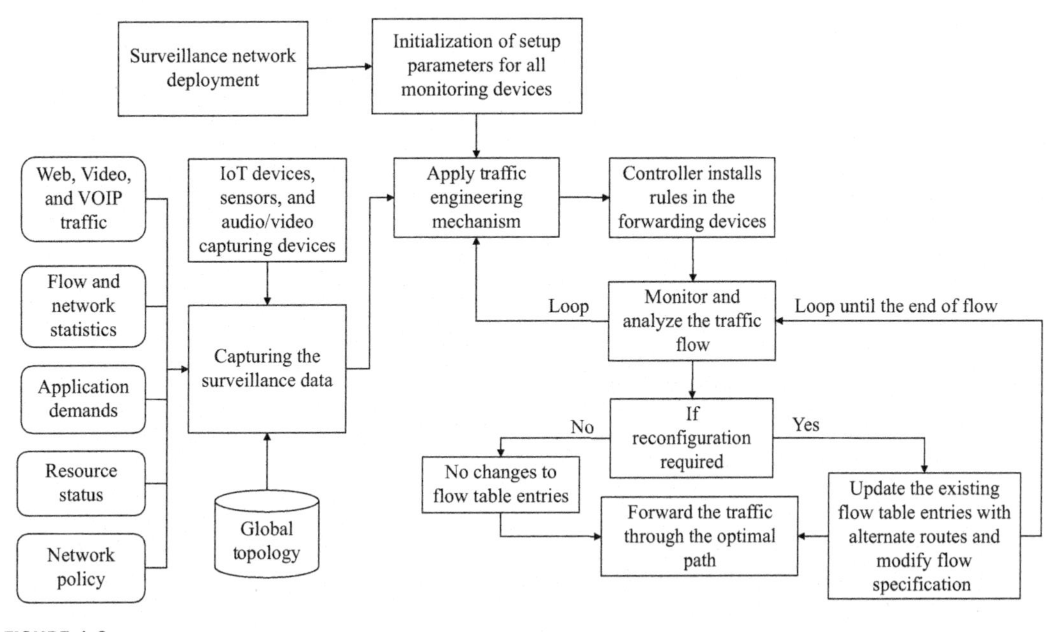


**FIGURE 1.6**

Block diagram of implementing TE mechanism. *TE*, Traffic engineering.

remote server and stored for further long-term analysis. Sometimes the captured recording is sent either to the local or remote server for an immediate evaluation of footage to take corrective actions on time. Therefore managing traffic flow plays a very important role in ensuring reliability.

Fig. 1.7 illustrates the traffic engineering framework for SDIoT. The controller has the capacity to perform traffic classification, flow routing, traffic policing, flow scheduling, load balancing, resource allocation, and energy management. Based on the control decisions made by the controller, the surveillance devices are programmed dynamically. The new and updated policies are installed into the devices by the controller.

Since a wide range of traffic profiles in dynamic is part of the network, a more accurate traffic classification based on flow features is required to guarantee the quality of service [23]. The SDN controller with traffic classification algorithm helps in traffic flow characterization and applying appropriate routing mechanism.

When forwarding traffic is computed using the open shortest path first (OSPF) algorithm, the same path will be always used to route the traffic from the surveillance devices to the storage or analysis server even though there are other nonutilized links. So certain links are overutilized leading to congestion, whereas other links are not at all used. Therefore to choose an optimal path, SDN is combined with the OSPF protocol [24]. To achieve load balancing links, weights and the flow-splitting ratio of the surveillance devices are dynamically changed. The controller device can arbitrarily split the flows coming into the intermediate forwarding devices to share the load on other available links and this minimization link utilization.

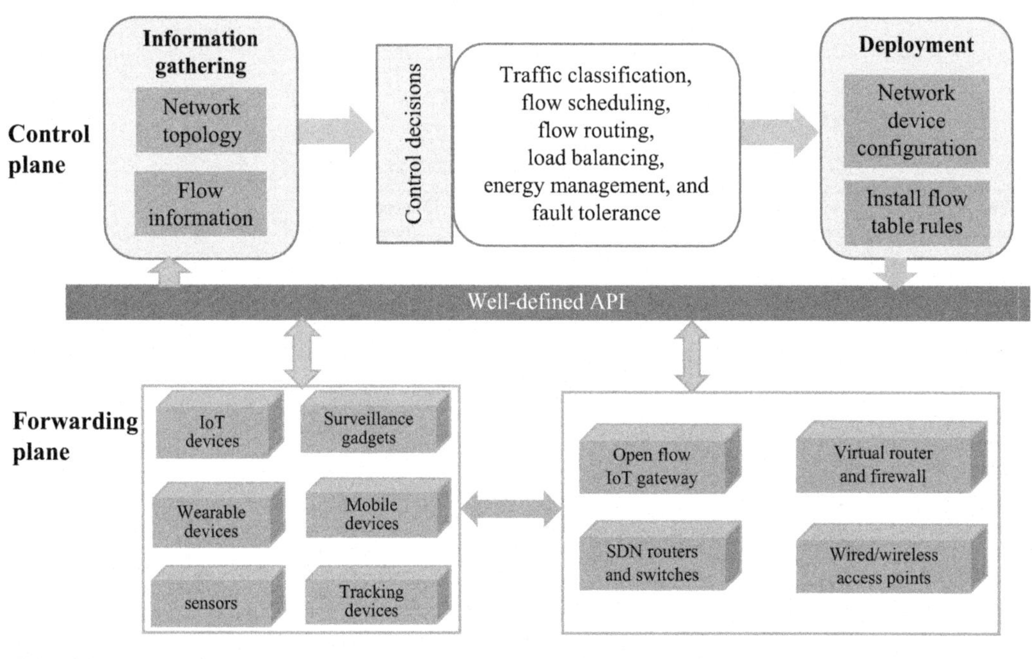


**FIGURE 1.7**

Traffic engineering in software defined networking with Internet of Things.

The two important routing metrics that are essential for routing traffic generated from multimedia applications are bandwidth and path delay. To ensure reliable communication of multimedia traffic across the devices in the entire surveillance network, the bandwidth constraints of the links have to be considered while computing the best route. It is recommended to always choose the path with the smallest delay route after eliminating all links with insufficient bandwidth [25]. The bottlenecked links are marked as critical and are not used until all the traffic in that link is delivered to the receiver. Links that have sufficient bandwidth is chosen for routing the traffic. Dynamic routing of traffic is possible with SDN implemented [26]. The newly arriving traffic flow should follow a route that does not interfere too much with routes that may be critical to satisfy future traffic demands. The path with the most available bandwidth is computed. If there is more than one such path, the path with less number of hops is considered optimal. It tries to load network links evenly such that no links are overloaded with too many traffic.

As modern surveillance consists of many IoT devices and low-power and computational capability devices with limited resources; the traffic engineering principles are enabled on the controller. The controller analyzes the traffic flow and dynamically controls the routes by predicting the flows types [27]. The predictable flows are sent by finding the path with sufficient bandwidth. The unpredictable flows follow the load balancing method, where the traffic is sent over multiple links so that links are not congested. To handle huge volumes of traffic, flows are allowed to pass through at least one SDN enabled devices such that the controller will have control over the devices [28] and flow routing is enhanced. This method reduces the maximum link utilization and minimizes packet loss and latency.

The controller can leverage the concept of segment routing (SR) [29] by dictating the behavior of the traffic flow. It works similar to source routing wherein the list of the intermediate devices through which the traffic has to forwarded is listed as segment identifier (SID) into the packet by the controller. The SR can be still improved by including the SID of the devices along a path that has sufficient bandwidth [30].

To enhance load balancing feature in surveillance system, the adaptive multipath techniques can be used [31]. The availability of the link and its bandwidth is used for path selection. Usually flows arrive at different times from several devices across the network. When the traffic flows arrive sequentially to avoid congestion, the controller calculates the traffic splitting ratio to find the consumed bandwidth of the link. The bandwidth is updated for the next round of flow processing [32]. Thus flows are forwarded incrementally and congestions are avoided. This method of forwarding results in high throughput, better load balancing, and minimizing link utilization.

As the surveillance system consists of low-power devices, it is essential to consider the energy consumption of the devices to provide reliable services. Since the surveillance tracking system has heterogeneous devices, each of them has different energy capacity. It is required of the controller to have knowledge of the device energy consumption status. The device should have at least the minimum energy level to become part of the network. The device with sufficient energy can actively involve in the transmission of the data. Because the information has to be sent reliably across the network, the controller has to ensure the energy level of each to become part of the network [33].

Fault tolerance is another important property of any network. Traffic engineering helps to reduce service degradation that occurs due to congestion and failure, for example, link or node failure. Implementing traffic engineering ensures that if a failure exists in the network, still the requested data can be delivered to the destination via alternate available routes. To address the single point of failure of the controller, distributed multiple controllers are used [34]. To dynamically perform traffic recovery, traffic engineering is applied in multidomain networks. SR scheme dynamically recovers traffic flows disrupted by link or node failures by minimizing the depth of the required segment list. Merging the protected path with a backup path as close as possible to the failure point reduces signaling overhead [35]. The best solution to guarantee fast recovery is rerouting disrupted traffic from the failed node.

To achieve the best result in terms of performance, the software defined traffic engineering mechanism is integrated along with IoT for the surveillance tracking system. This improves significantly the traffic management by improving throughput, link utilization, reducing congestion, and drop loss. It improves the robustness of the network by reducing service degradation and building a most reliable surveillance tracking system.

## 1.6 CONCLUSION

The surveillance tracking system has gained momentum in recent years because of the wide prospects of its usage in various application fields. In this paper, the surveillance tracking system has been explored to enhance its working operation. As the traditional network architecture is not flexible to handle the huge volume of surveillance video traffic and dynamic traffic demands, the SDN technology is integrated along with IoT. The novel framework for SDIoT was proposed, which will help to handle unpredictable huge flow of traffic from multiple heterogeneous devices. The traffic engineering principles support dynamic routing of traffic for improving performance in terms of link bandwidth utilization, energy consumption, and optimal path selection. For future work, this framework will be further investigated to implement the surveillance tracking system for multiple potential applications.

## REFERENCES

[1] Y. Yin, Y. Zeng, X. Chen, Y. Fan, The Internet of Things in healthcare: an overview, J. Ind. Inf. Integr. 1 (2016) 3–13.

[2] N. Haering, P.L. Venetianer, A. Lipton, The evolution of video surveillance: an overview, Mach. Vis. Appl. 19 (5–6) (2008) 279–290.

[3] M. Alaa, A.A. Zaidan, B.B. Zaidan, M. Talal, M.L.M. Kiah, A review of smart home applications based on Internet of Things, J. Netw. Comput. Appl. 97 (2017) 48–65.

[4] R. Kunst, L. Avila, E. Pignaton, S. Bampi, J. Rochol, Improving network resources allocation in smart cities video surveillance, Comput. Netw. 134 (2018) 228–244.

[5] A.H. Alavi, P. Jiao, W.G. Buttlar, N. Lajnef, Internet of Things-enabled smart cities: state-of-the-art and future trends, Measurement 129 (2018) 589–606.

[6] P. Remagnino, S.A. Velastin, G.L. Foresti, M. Trivedi, Novel concepts and challenges for the next generation of video surveillance systems, Mach. Vis. Appl. 18 (3–4) (2007) 135–137.

[7] P.K. Atrey, N.C. Maddage, M.S. Kankanhalli, Audio based event detection for multimedia surveillance, Event (Lond.) (2006) 813–816.

[8] B. Uzkent, B.D. Barkana, "Pitch-range based feature extraction for audio surveillance systems," in: IEEE Computer Society, 8th International Conference on Information Technology: New Generations, 2011, 476–480.
[9] V. Tsakanikas, T. Dagiuklas, Video surveillance systems-current status and future trends R, Comput. Electr. Eng. 70 (2018) 736–753.
[10] A.H. Meghdadi, P. Irani, Interactive exploration of surveillance video through action shot summarization and trajectory visualization, IEEE Trans. Vis. Comput. Graph. 19 (12) (2013) 2119–2128.
[11] M. Guennoun, S. Khattak, B. Kapralos, K. El-khatib, Augmented Reality-Based Audio/Visual Surveillance System, in: IEEE International Workshop on Haptic Audio visual Environments and Games, 2008, pp. 18–19.
[12] D. Potoglou, F. Dunkerley, S. Patil, N. Robinson, Public preferences for Internet surveillance, data retention and privacy enhancing services: evidence from a Pan-European study, Comput. Hum. Behav. 75 (2017) 811–825.
[13] K. Michael, R. Clarke, Location and tracking of mobile devices: Überveillance stalks the streets, Comput. L. Sec. Rev. 29 (3) (2013) 216–228.
[14] L. Pellegrin, C. Vignal, J. Meynard, X. Deparis, H. Chaudet, Dealing with uncertainty when using a surveillance system, Int. J. Med. Inform. 104 (2017) 65–73.
[15] M. Lakshminarasimhan, Advanced Traffic Management System Using Internet of Things, Researchgate, 2016.
[16] G. Guido, A. Vitale, F.F. Saccomanno, V. Astarita, Vehicle tracking system based on videotaping data, Procedia Soc. Behav. Sci. 111 (2014) 1123–1132.
[17] S. Kannimuthu, C.K. Somesh, P.D. Mahendhiran, D. Bhanu, K.S. Bhuvaneshwari, Certain investigation on significance of Internet of Things (IoT) and Big Data in vehicle tracking system, Indian J. Sci. Technol. 9 (39) (2016).
[18] V.V. Prasad, C. Sekhar, C. Naveen, R. Vishal, Object's action detection using GMM algorithm for smart visual surveillance system, Procedia Comp. Sci. 133 (2018) 276–283.
[19] D. Oosterlinck, D.F. Benoit, P. Baecke, N. Van de Weghe, Bluetooth tracking of humans in an indoor environment: an application to shopping mall visits, Appl. Geogr. 78 (2017) 55–65.
[20] S. Agarwal, Traffic engineering in software defined networks, in: Proceedings IEEE INFOCOM 2013, 2211–2219.
[21] I. Akyildiz, A. Lee, P. Wang, M. Luo, W. Chou, Research challenges for traffic engineering in software defined networks, IEEE Netw. (2016) 52–58.
[22] S. Bera, S. Misra, A.V. Vasilakos, Software-defined networking for Internet of Things: a survey, IEEE Internet Things J. 4 (6) (2017) 1994–2008.
[23] A.S. Da Silva, C.C. Machado, R.V. Bisol, L.Z. Granville, A. Schaeffer-Filho, Identification and selection of flow features for accurate traffic classification in SDN, in: IEEE 14th International Symposium on Network Computing and Applications, 2015, 134–141.
[24] Y. Guo, Z. Wang, X. Yin, X. Shi, J. Wu, Traffic engineering in SDN/OSPF hybrid network, in: IEEE 22nd International Conference on Network Protocols, 2014, 563–568.
[25] S. Tomovic, N. Lekic, I. Radusinovic, A new approach to dynamic routing in SDN networks, in: 18th Mediterranean Electrotechnical Conference, 2016.
[26] A. Karaman, Constraint-based routing in traffic engineering, in: International Symposium on Computer Networks, 2006.
[27] Y. Takahashi, K. Ishibashi, M. Tsujino, N. Kamiyama, K. Shiomoto, T. Otoshi, et al., Separating predictable and unpredictable flows via dynamic flow mining for effective traffic engineering, in: IEEE International Conference on Communications, 2016.

[28] C. Ren, S. Wang, J. Ren, X. Wang, T. Song, D. Zhang, Enhancing traffic engineering performance and flow manageability in Hybrid SDN, in: IEEE Global Communications Conference, 2016.
[29] L. Davoli, L. Veltri, P.L. Ventre, G. Siracusano, S. Salsano, Traffic engineering with segment routing: SDN-based architectural design and open source implementation, in: Proceedings of the European Workshop on Software Defined Networks, EWSDN, vol. 2015, 2015, pp. 111–112.
[30] E. Moreno, A. Beghelli, F. Cugini, Traffic engineering in segment routing networks, Comput. Netw. 114 (2017) 23–31.
[31] S. Sahhaf, W. Tavernier, D. Colle, M. Pickavet, Adaptive and reliable multipath provisioning for media transfer in SDN-based overlay networks, Comput. Commun. 106 (2017) 107–116.
[32] W. Wang, W. He, J. Su, Enhancing the effectiveness of traffic engineering in hybrid SDN, in: IEEE International Conference on Communications, 2017.
[33] C. Kharkongor, T. Chithralekha, R. Varghese, A SDN controller with energy efficient routing in the Internet of Things (IoT), Procedia Comput. Sci. 89 (2016) 218–227.
[34] Y.E. Oktian, S.G. Lee, H.J. Lee, J.H. Lam, Distributed SDN controller system: a survey on design choice, Comput. Netw. 121 (2017) 100–111.
[35] A. Giorgetti, A. Sgambelluri, F. Paolucci, F. Cugini, P. Castoldi, Segment routing for effective recovery and multi-domain traffic engineering, J. Opt. Commun. Netw. 9 (2) (2017) A223.

CHAPTER

# AN EFFICIENT PROVABLY SECURE IDENTITY-BASED AUTHENTICATED KEY AGREEMENT SCHEME FOR INTERVEHICULAR AD HOC NETWORKS

2

**Renu Mary Daniel[1], Elijah Blessing Rajsingh[1], Salaja Silas[1] and Sharmila Anand John Francis[2]**
*[1]Department of Computer Science and Engineering, Karunya Institute of Technology and Sciences, Coimbatore, India [2]Department of Computer Science, King Khalid University, Abha, Saudi Arabia*

## 2.1 INTRODUCTION

Vehicular ad hoc networks (VANETs) are comprised of moving vehicles and stationary roadside units (RSUs) equipped with sophisticated sensors and dedicated short-range wireless communication devices. The vehicles can interchange information among themselves, called vehicle-to-vehicle (V2V) communication, or with the roadside infrastructure elements like RSUs, called vehicle-to-infrastructure (V2I) communication. Intervehicular (V2V) communications are of particular interest, since for critical road safety applications, real-time information about potential dangers, such as weather conditions, road conditions, or traffic congestion, can be communicated among the vehicles. Moreover, advanced safety alerts like electronic brake light warning, lane change warning, blind spot warning, forward collision warning, or control loss warning can further assist in reducing road traffic crashes. In each of these scenarios, vehicles share basic information such as location coordinates, direction of travel, speed, braking status, and loss of stability. It is essential to maintain a secure authenticated channel for communication among vehicles, since authentic information can assist traffic control units to make critical decisions in dispensing emergency services with a minimal delay. Even for applications like truck platooning, maliciously inserted data, like a wrong braking status, can lead to collisions.

As for comfort applications, V2V can facilitate file sharing, message passing, Internet access, and even multiplayer network games. Ensuring data privacy is a prime concern in such applications, and it is imperative to establish secure authenticated channels to effectively utilize these applications to enhance travel comfort. To this end, we propose an identity-based authenticated key agreement (ID-AKA) protocol for establishing a shared secret key between two authentic entities, in a VANET. The system architecture is comprised of a trusted authority called the key generation center (KGC), vehicles with on-board units (OBUs), and RSUs. The general architecture of a VANET and few of its applications are illustrated in Fig. 2.1.

The Cognitive Approach in Cloud Computing and Internet of Things Technologies for Surveillance Tracking Systems.
DOI: https://doi.org/10.1016/B978-0-12-816385-6.00002-7

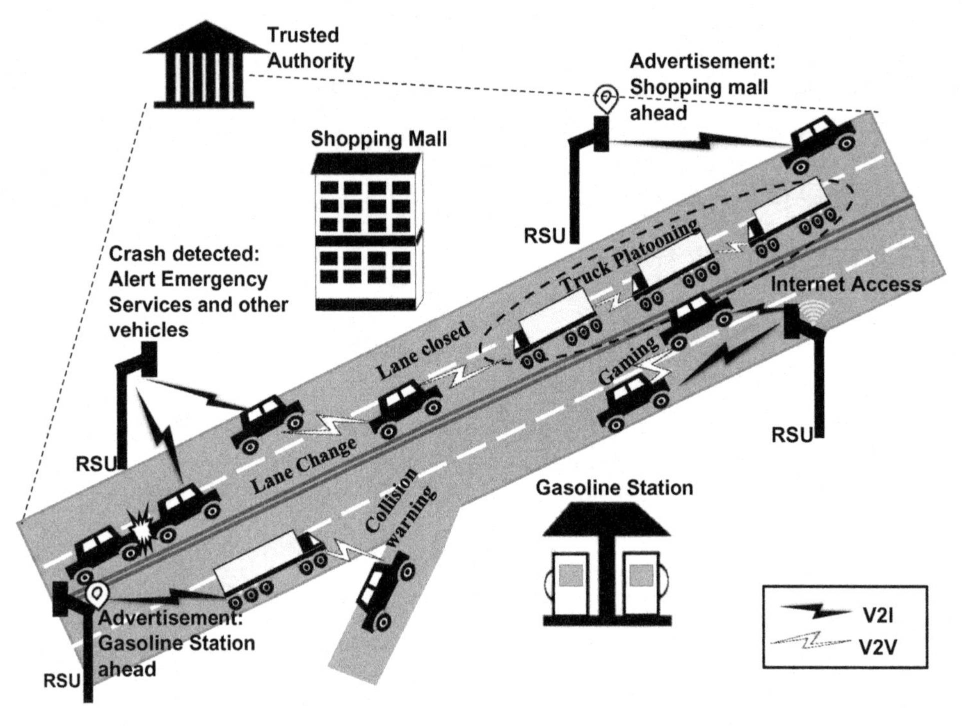


**FIGURE 2.1**

General architecture of a vehicular ad hoc network and its applications.

The trusted authority, which also serves as the KGC, registers each vehicle (offline) after validating the identity and license of the driver and generates the unique vehicle registration number. Since the registration number is unique within a country (in most cases), it can serve as the vehicle's pseudonym identity. After registration, the KGC issues the private key corresponding to the vehicle's identity. The private key along with the system public parameters will be incorporated into the memory unit of the vehicle's OBU. Furthermore, the embedded microcontroller in the OBU performs various computations including cryptographic operations.

### 2.1.1 RELATED WORK

The initial key agreement protocols in VANETs ensured public key authenticity using the public key infrastructure (PKI) model. The implication is that each vehicle or RSU must obtain the valid certificate of its peer, perform certificate verification, and then negotiate a basic Diffie–Hellman session key to communicate among each other or to obtain value-added services from various service providers [1–3]. To prevent vehicle tracking by malicious entities and to preserve anonymity while accessing services, several existing protocols require frequent pseudonym changes. This requirement further increases the computational and storage overhead of the system, since the pseudonym-related certificate of each entity must be updated at regular intervals [4,5]. To eliminate the need for generating numerous certificates, Li et al. [6] proposed a key agreement scheme for intervehicular communications, using blind signatures. However, the protocol is deemed to be inefficient, since the underlying cyclic group requires a composite modulus comprising of four primes. Studies prove that the size of the composite modulus must be at least 4096 bits, to prevent factoring attacks [7,8]. Subsequently, several protocols were designed to facilitate RSU-aided distributed certificate generation services [1,9,10], to reduce the workload of the central certificate authority. These protocols require expensive bilinear pairing computations for certificate generation, initial vehicle registration, and/or key agreement, rendering them inefficient. Furthermore, the partially identity-based key agreement schemes in [11] and [12] require bilinear pairings for vehicle registration and symmetric key establishment, respectively. It is to be noted that none of these protocols are analyzed in a formal security model. Apparently, a heuristic analysis is insufficient to guarantee security against a wide range of attacks such as basic impersonation (BI) attack, key compromise impersonation (KCI) attack, unknown key share (UKS) attack, lack of perfect forward secrecy (PFS), and lack of ephemeral secret key leakage (ESKL) resilience.

Recently, Dang et al. (DXC) [13] proposed an efficient extended CK (eCK) [14] secure ID-AKA scheme for securing V2V communications, based on the gap Diffie–Hellman (GDH) assumption. The authors claim that the protocol ensures provable security against BI, KCI, and UKS attacks. However, a cryptanalysis of the protocol reveals several flaws in the security proof. We illustrate how an active adversary can exploit these vulnerabilities to unleash a successful KCI attack. Subsequently, we propose an efficient eCK secure ID-AKA protocol, based on the GDH assumption, with provable security against all the aforementioned attacks. As in [13], the proposed scheme does not provide anonymity. The intuition is that the critical information being interchanged in a V2V communication is acceptable, only if the sender can be authenticated. Furthermore, the ring signature generation and verification in anonymous schemes increase the computational complexity of these highly dynamic, short-term intervehicular communications. Moreover, since the peers establish a secure authenticated channel by negotiating a session key, the messages exchanged between them remain unintelligible to external adversaries. Even if an attacker obtains the identities of the communicating peers, the session key cannot be computed, unless the adversary additionally acquires both the static private key and the ephemeral private key of at least one of the communicating peers.

The rest of the chapter is organized as follows. In Section 2.2, we discuss the hardness assumptions used for the protocol design as well as the desirable security attributes of AKA protocols. Section 2.3 provides a brief overview about the general adversarial capabilities and simulation of the eCK security model. The proposed ID-AKA protocol is presented in Section 2.4, and the security proof is provided in Section 2.5. Section 2.6 illustrates the security vulnerabilities in the DXC's ID-AKA scheme. Section 2.7 provides a comparative analysis of the security and efficiency of the proposed scheme with other ID-AKA protocols. The concluding remarks are provided in Section 2.8.

## 2.2 PRELIMINARIES

This section introduces the basic hardness assumptions used for the protocol design as well as the desirable security attributes of AKA schemes.

### 2.2.1 HARDNESS ASSUMPTIONS

For a given security parameter $\lambda$, let $G$ be a cyclic additive group of prime order $q$, defined by a point addition operation over an elliptic curve $E(F_p)$. Let $P$ be the generator of $G$. We use the notation $\in_R$ to denote "chosen uniformly at random."

- *Computational Diffie–Hellman (CDH) problem*: For unknown values $x, y \in_R Z_q^*$, given the tuple $(P, X = xP, Y = yP)$, no probabilistic polynomial time (PPT) adversary can compute $xyP$.
- *Decisional Diffie–Hellman (DDH) problem*: For unknown values $x, y, c \in_R Z_q^*$, given the tuple $(P, X = xP, Y = yP, C = cP)$, no PPT adversary determines whether $xyP = cP$.
- *GDH problem*: Consider a DDH oracle $DDH(*, *, *)$, which, on input $(X = xP, Y = yP, C = cP)$, returns 1 if $xyP = cP$ and 0 otherwise. No PPT adversary can solve the CDH problem to compute $xyP$, even if it is additionally provided with a DDH oracle.

Let $DLOG(): G \to Z_q^*$ denote a function that maps a point $xP \in G$ to its discrete logarithm value $x \in Z_q^*$.

### 2.2.2 DESIRABLE SECURITY ATTRIBUTES OF AUTHENTICATED KEY AGREEMENT PROTOCOLS

AKA protocols are required to satisfy the following security criteria.

- *Known key security (KKS)*: The protocol generates a unique session key for each session. An adversary in possession of session keys related to other sessions cannot recover the current session key.
- *No key control*: Each participating entity equally contributes information toward the creation of the shared key. Neither of the participants can force the session key in part or in its entirety upon its peer.
- *BI attack resilience*: An adversary cannot impersonate a legitimate user, unless it obtains the latter's static private key.
- *KCI resilience*: An adversary in possession of an entity's static private key cannot impersonate other uncompromised users in the system, to the compromised entity.
- *Weak perfect forward secrecy (wPFS)*: Any adversary that has not actively participated in the generation of the ephemeral private keys corresponding to a previously established session cannot recover the session key, even if it compromises the static private keys of both the communicating peers of that session.
- *ESKL resilience*: An adversary in possession of both the ephemeral private keys corresponding to a session cannot compute the session key, unless it additionally compromises the static private key of at least one of the concerned peers.
- *UKS attack resilience*: An entity cannot be coerced into thinking that it is sharing the key with an adversary, when the key is actually being shared with another honest peer.

## 2.3 SECURITY MODEL

AKA protocols are deemed to be suitable for practical applications, only if they are proven secure in a formal security model. Bellare and Rogaway (BR) [15] proposed the first formal security model for AKA protocols based on the indistinguishability game. The BR model and its variants [16], [17] capture KKS; however, these models fail to capture KCI attacks, ESKL attacks, and wPFS property. Canetti and Krawczyk (CK) [18] proposed the CK security model that empowers the adversary with session state reveal and corrupt queries. Nevertheless, since these queries cannot be issued against the test session, the CK model fails to capture ESKL and KCI attacks [19]. In order to ensure provable security against KCI and ESKL attacks, Lamacchia et al. [20] proposed the eCK model that allows the adversary to issue any nontrivial combination of static private key reveal queries, as well as ephemeral private key reveal queries, even for the test session. Therefore, to ensure a maximum-exposure-resilience property, the proposed protocol is analyzed based on the eCK security model. A brief overview of the eCK security model is provided below.

### 2.3.1 PARTICIPANTS

The participating entities are defined by a finite set $U$ of fixed size $n$. Each participating peer $u_i$ with an identity $ID_i$, static private (public) key $d_i(P_i)$, and ephemeral private (public) key $t_i(T_i, \text{respectively})$ is modeled as a PPT machine. Each peer can simultaneously execute multiple instances (sessions) of the protocol, in parallel. Each entity can be activated by any one of the following messages: (1) $(I, ID_i, ID_j)$; or (2) $(R, ID_i, ID_j, M_i)$. If the entity $ID_i$ is activated by the first message, it takes up the role of the session initiator; otherwise, it acts as the responding peer.

### 2.3.2 SESSION

Each session is defined by the session identifier $\Pi = (role, ID_i, ID_j, M_i, M_j)$, where *role* is the role (initiator $I$ or responder $R$) of the session owner, $ID_i$ is the identity of the session owner, $ID_j$ is the identity of the peer, $M_i$ is the message sent by $ID_i$, and $M_j$ is the message received by $ID_i$. Two sessions $\Pi = (role, ID_i, ID_j, M_i, M_j)$ and $\Pi^* = (role*, ID_j^*, ID_i^*, M_j^*, M_i^*)$ are said to be matching, if $role \neq role*$, $ID_i = ID_i^*, ID_j = ID_j^*$, $M_i = M_i^*$, and $M_j = M_j^*$. For instance, two sessions $\Pi = (I, ID_i, ID_j, (R_i, T_i), (R_j, T_j))$ and $\Pi^* = (R, ID_j, ID_i, (R_j, T_j), (R_i, T_i))$ are matching, where $I(R)$ denotes the roles initiator (responder, respectively), tuple $(R_i, T_i)$ corresponds to the message $M_i(\text{or } M_i^*)$, and tuple $(R_j, T_j)$ corresponds to the message $M_j(\text{or } M_j^*)$, respectively.

### 2.3.3 ADVERSARY

The eCK model considers a PPT adversary $\mathscr{A}$, which is capable of controlling all the communications in the session, including the session activations, using the following adversarial queries.

- *Send(Π, message)*: The adversary $\mathscr{A}$ can activate the initiating peer $ID_i$ by the message $(I, ID_i, ID_j)$, or the responding peer $ID_j$ by the message $(R, ID_j, ID_i, (R_i, T_i))$. In addition, $\mathscr{A}$ can also send the message $(I, ID_i, ID_j, (R_j, T_j))$ back to $ID_i$, as the response from $ID_j$. In all these cases, the concerned peers act according to the protocol specification.

  Furthermore, to simulate the leakage of secret data in a session, the adversary is allowed to issue the following queries.
- *SessionKeyReveal*($\Pi$): $\mathscr{A}$ is provided access to the session key corresponding to the completed session $\Pi$.

- *EphemeralPrivateKeyReveal*($\Pi$): $\mathscr{A}$ is provided access to the ephemeral key generated by the owner of the session $\Pi$.
- *StaticPrivateKeyReveal*($u_i$): $\mathscr{A}$ is provided access to the static private key of the participating entity $u_i$.
- *CreatePeer*($u_i$): $\mathscr{A}$ registers an entity $u_i$, with an identity $ID_i$. The static public–private key pair $(d_i, P_i)$ of the entity $u_i$ is generated according to the protocol specification. These values are returned to $\mathscr{A}$. The entity $u_i$ registered by the *CreatePeer*($u_i$) query is completely controlled by $\mathscr{A}$ and is said to be dishonest.

### 2.3.4 FRESH SESSION

Let $\Pi$ denote a completed session between two honest peers $(u_i, u_j)$, where $u_i$ is the initiating peer and $u_j$ is the responder. Let $\Pi^*$ denote the matching session of $\Pi$ (if it exists). The session $\Pi$ is said to be locally exposed, if any of the following conditions hold:

- $\mathscr{A}$ issues the *SessionKeyReveal*($\Pi$) query.
- $\mathscr{A}$ issues both *StaticPrivateKeyReveal*($u_i$) and *EphemeralPrivateKeyReveal*($\Pi$).

The session $\Pi$ is said to be exposed, if: (1) $\Pi$ is locally exposed; (2) the matching session $\Pi^*$ exists and is locally exposed; or (3) $\Pi^*$ does not exist and $\mathscr{A}$ issues a *StaticPrivateKeyReveal*($u_j$) query. The session $\Pi$ is fresh if none of these conditions hold and $\Pi$ remains unexposed.

### 2.3.5 SECURITY EXPERIMENT

The security experiment is defined as an adversarial game between a challenger $\mathscr{C}$ that tries to solve the GDH assumption, using an adversary $\mathscr{A}$ that tries to compromise the security of the ID-AKA protocol. Initially, $\mathscr{A}$ is provided with the finite set of honest entities $U = \{u_1, \ldots, u_n\}$. $\mathscr{A}$ can issue a polynomial number of send queries or reveal queries, in any order. Once the query phase is over, $\mathscr{A}$ issues the *Test*($\Pi$) query during the challenge phase. The *Test*($\Pi$) query is characterized by the following features.

*Test*($\Pi$): The adversary $\mathscr{A}$ can issue only a single query of this form. The session $\Pi$ associated with the test query must be fresh. The challenger $\mathscr{C}$ responds to the test query by choosing a random bit $b \in_R \{0, 1\}$. If $b = 0$, then $\mathscr{C}$ returns the session key generated by $\Pi$. Otherwise, a randomly chosen key is returned to $\mathscr{A}$.

The adversary continues the query phase and finally makes a guess $b'$. $\mathscr{A}$ wins the game if the freshness of $\Pi$ is still preserved and the guess bit $b' = b$. The advantage of $\mathscr{A}$ in the security experiment is defined as $Adv_{\Pi}^{AKA}(\mathscr{A}) = \left|\Pr\left(b' = b\right) - \frac{1}{2}\right|$.

### 2.3.6 DEFINITION 1 (ECK SECURITY OF IDENTITY-BASED AUTHENTICATED KEY AGREEMENT PROTOCOL)

An ID-AKA protocol is said to be secure in the eCK security model, if the following conditions are satisfied.

- In the presence of a passive adversary that faithfully conveys the protocol messages, matching sessions compute the same session key.
- For any PPT adversary $\mathscr{A}$, $Adv_{\Pi}^{AKA}(\mathscr{A})$ is a negligible function of $\lambda$.

## 2.4 PROVABLY SECURE IDENTITY-BASED AUTHENTICATED KEY AGREEMENT PROTOCOL FOR V2V COMMUNICATIONS

The proposed ID-AKA protocol for securing V2V communications is comprised of three phases, namely the setup phase, the entity registration phase, and the key agreement phase.

### 2.4.1 SETUP PHASE

Let $\lambda$ be the security parameter. Let $G$ be an elliptic curve group of prime order $q$ and let the point $P$ be the generator of the additive group $G$. Let $G^*$ denote the set of nonidentity elements in $G$. The KGC chooses the master secret key $s \in_R z_q^*$ and computes the corresponding master public key $P_{pub} = sP$. The KGC also defines the hash functions: $H_1$:$\{0,1\}^* \times G \rightarrow Z_q^*$ and $H_2$:$\{0,1\}^* \times \{0,1\}^* \times G^7 \rightarrow \{0,1\}^z$, where $z$ denotes the size of the final session key. The system public parameters are defined by the tuple $\{q, G, P, H_1, H_2\}$.

### 2.4.2 ENTITY REGISTRATION PHASE

For each vehicle $u_i$, with the identity $ID_i$, the KGC chooses a randomizing element $r_i \in_R z_q^*$ and computes a Schnorr [21] signature $d_i = r_i + H_1(ID_i, R_i)s$, where $R_i = r_iP$. The value $d_i$ corresponds to the static private key of the entity $u_i$. Given the entity's public elements $(ID_i, R_i)$, the corresponding static public key can be computed as $P_i = R_i + H_1(ID_i, R_i)P_{pub} = d_iP$. The static private key is stored in the tamper-proof device of the OBU. The system public parameters as well as the master public key are also incorporated into the vehicle's OBU.

### 2.4.3 KEY AGREEMENT PHASE

The key agreement phase between the vehicles $u_A$ and $u_B$ with the identities $ID_A$ and $ID_B$ proceeds as shown in Fig. 2.2.

1. The initiating peer $u_A$, upon activation by the message $(I, ID_A, ID_B)$, chooses the ephemeral private key $t_A \in_R Z_q^*$, computes the corresponding ephemeral public key $T_A = t_AP$, and sends the tuple $(R, ID_B, ID_A, (R_A, T_A))$ to the user $u_B$. The entity $u_A$ also creates a session defined by $\Pi = (I, ID_A, ID_B, (R_A, T_A), -)$.
2. On receiving the message $(R, ID_B, ID_A, (R_A, T_A))$ from $u_A$, the responding peer $u_B$ checks if $T_A \in G^*$; if so, it chooses the ephemeral private key $t_B \in_R Z_q^*$, computes $T_B = t_BP$, and sends the tuple $(I, ID_A, ID_B, (R_B, T_B))$ to the session initiator $u_A$. Concurrently, the peer $u_B$ defines $\Pi^* = (R, ID_B, ID_A, (R_B, T_B), (R_A, T_A))$. The entity $u_B$ then computes the shared secrets $K_{BA_1} = (t_B + 2d_B)(T_A + 2P_A)$, $K_{BA_2} = (t_B - d_B)(T_A - P_A)$, and $K_{BA_3} = (2t_B + d_B)(2T_A + P_A)$,

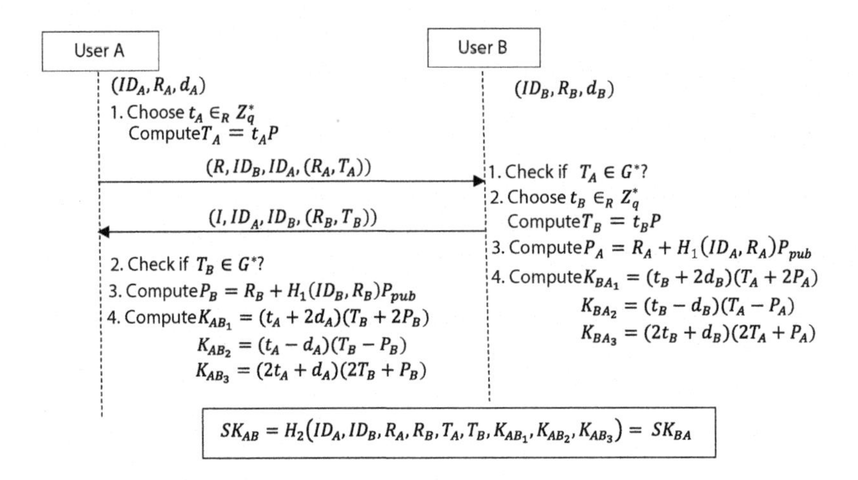


**FIGURE 2.2**

Key agreement phase in the proposed identity-based authenticated key agreement protocol.

where $P_A = R_A + H_1(ID_A, R_A)P_{pub}$ is the static public key of $u_A$. Finally, the session key is computed as $SK_{BA} = H_2(ID_A, ID_B, R_A, R_B, T_A, T_B, K_{BA_1}, K_{BA_2}, K_{BA_3})$.

**3.** On receiving the message $(I, ID_A, ID_B, (R_B, T_B))$, the entity $u_A$ checks if $T_B \in G^*$ and there exists an incomplete session of the form $\Pi = (I, ID_A, ID_B, (R_A, T_A), -)$, then it completes the tuple as $\Pi = (I, ID_A, ID_B, (R_A, T_A), (R_B, T_B))$ and computes the static public key of peer $u_B$ as $P_B = R_B + H_1(ID_B, R_B)P_{pub}$. Subsequently, it computes the shared secrets $K_{AB_1} = (t_A + 2d_A)(T_B + 2P_B)$, $K_{AB_2} = (t_A - d_A)(T_B - P_B)$, and $K_{AB_3} = (2t_A + d_A)(2T_B + P_B)$. The resultant session key is derived as $SK_{AB} = H_2(ID_A, ID_B, R_A, R_B, T_A, T_B, K_{AB_1}, K_{AB_2}, K_{AB_3})$.

The correctness of the session key can be verified as follows: $K_{AB_1} = (t_A + 2d_A)(T_B + 2P_B) = (t_A + 2d_A)(t_B + 2d_B)P = (t_B + 2d_B)(T_A + 2P_A) = K_{BA_1}$, $K_{AB_2} = (t_A - d_A)(T_B - P_B) = (t_A - d_A)(t_B - d_B)P = (t_B - d_B)(T_A - P_A) = K_{BA_2}$, and $K_{AB_3} = (2t_A + d_A)(2T_B + P_B) = (2t_A + d_A)(2t_B + d_B)P = (2t_B + d_B)(2T_A + P_A) = K_{BA_3}$.

## 2.5 SECURITY ANALYSIS

**Theorem 1::** *Assuming that the GDH problem is intractable in $(G, P)$ and considering the hash functions $(H_1, H_2)$ as random oracles, the proposed ID-AKA protocol for V2V communications is secure in the eCK model.*

**Proof::** Consider that a PPT adversary $\mathscr{A}$ can win the security game, as illustrated in Section 2.3, by distinguishing the session key of the fresh session in the test query, from a random key, with the probability $\frac{1}{2} + \nu(\lambda)$, such that $\nu(\lambda)$ is non-negligible. Then, we prove that it is possible to construct a GDH solver $\mathscr{C}$ using $\mathscr{A}$, also with a nonnegligible probability. Let $(X, Y)$ be a CDH instance, such that $\mathscr{C}$ is unaware of the values $(x = DLOG(X), y = DLOG(Y))$. The GDH solver $\mathscr{C}$ is provided with a $DDH(*, *, *)$ oracle and simulates the security game for the adversary $\mathscr{A}$. $\mathscr{C}$ simulates the KGC and generates the system public parameters, along with the master public–private key pair $(s, P_{pub} = sP)$. The challenger $\mathscr{C}$ then creates $n$ honest entities, by generating their static public–private key pairs according to the protocol specification. Let $m$ denote the maximum admissible number of protocol sessions. Let $\Pi = (I, ID_A, ID_B, (R_A, T_A), (R_B, T_B))$ denote the test session, which is a completed session, executed between two honest peers $(u_A, u_B)$. Let $F$ denote the event that $\mathscr{A}$ succeeds in the distinguishing game. Let $QH_2$ denote the event that $\mathscr{A}$ issues the $H_2$ oracle query $H_2(ID_A, ID_B, R_A, R_B, T_A, T_B, K_{AB_1}, K_{AB_2}, K_{AB_3})$, concerning the test session $\Pi$. Let $\overline{QH_2}$ denote the complement of the event $QH_2$. It is easy to see that, if the event $QH_2$ does not occur, the probability of $\mathscr{A}$ succeeding the security game is no better than a random coin flip. Hence, $\Pr(F \wedge \overline{QH_2}) \leq \frac{1}{2}$, and $\Pr(F) = \Pr(F \wedge \overline{QH_2}) + \Pr(F \wedge QH_2) \leq \frac{1}{2} + \Pr(F \wedge QH_2)$. The implication is that, for the adversary to win the game, $\Pr(F \wedge QH_2) \geq \nu(\lambda)$, and we denote the event $F \wedge QH_2$ as $W$. For the successful simulation of the distinguishing game, the freshness of $\Pi$ must be preserved. Therefore we consider the following complementary cases.

1. The matching session $\Pi^*$ corresponding to the test session $\Pi$ does not exist. In this case, to maintain the freshness of $\Pi$, the adversary cannot issue the *RevealStaticPrivateKey*$(ID_B)$ query and either of the following:
   a. $\mathscr{A}$ cannot issue the *EphemeralPrivateKeyReveal*$(\Pi)$ query—Event $F_{1a}$.
   b. $\mathscr{A}$ cannot issue the *StaticPrivateKeyReveal* $(ID_A)$ query—Event $F_{1b}$.
2. The matching session $\Pi^*$ corresponding to the test session $\Pi$ exists, and any of the following:
   a. $\mathscr{A}$ cannot issue either the *StaticPrivateKeyReveal*$(ID_A)$ query or *StaticPrivateKeyReveal* $(ID_B)$ query—Event $F_{2a}$.
   b. $\mathscr{A}$ cannot issue either the *StaticPrivateKeyReveal*$(ID_A)$ query or *EphemeralPrivateKeyReveal*$(\Pi^*)$ query—Event $F_{2b}$.
   c. $\mathscr{A}$ cannot issue either the *EphemeralPrivateKeyReveal*$(\Pi)$ query or *StaticPrivateKeyReveal* $(ID_B)$ query—Event $F_{2c}$.
   d. $\mathscr{A}$ cannot issue either the *EphemeralPrivateKeyReveal*$(\Pi)$ query or *EphemeralPrivateKeyReveal*$(\Pi^*)$ query—Event $F_{2d}$.

### 2.5.1 EVENT $W \wedge F_{1A}$

In this case, the matching session $\Pi^*$ does not exist; hence, for the test session, the incoming ephemeral private key $t_B$ originating from the responding peer $u_B$ is chosen by $\mathscr{A}$. The challenger $\mathscr{C}$ generates the static private key for each user $u_i \neq u_B$, by choosing random elements $(d_i, h_i) \in_R Z_q^*$ and setting $R_i = d_iP - h_iP_{Pub}$. Thus the static public key $P_i = d_iP = R_i + h_iP_{Pub}$, as per the protocol specification. The challenger then embeds the CDH instance $(X, Y)$ as follows: $\mathscr{C}$ randomly chooses two users $(u_A, u_B) \in_R u_{i...n}$ and a protocol instance $l \in_R [1, m]$, in an attempt to guess the test session $\Pi$, which is correct with a probability $\frac{1}{n^2m}$. $\mathscr{C}$ then sets the ephemeral private key of the test session as $T_A = X$. The static public key of the user $u_B$ is set as $P_B = Y$, by selecting $h_B \in_R Z_q^*$ and assigning $R_B = Y - h_BP_{pub}$. Thus $\mathscr{C}$ knows the static private keys of all the users in the system except $u_B$. The challenger $\mathscr{C}$ responds to the user activations and oracle queries issued by the adversary $\mathscr{A}$ as shown below.

#### 2.5.1.1 Simulation

$\mathscr{C}$ maintains two lists $L_{H_2}$ and $L_s$, to keep track of the $H_2$ oracle queries and the *SessionKeyReveal* queries, respectively. Whenever $\mathscr{A}$ activates protocol participants via the send queries, $\mathscr{C}$ responds in the conventional manner.

1. *Send*$(I, ID_i, ID_j)$: $\mathscr{C}$ chooses $t_i \in_R Z_q^*$, computes $T_i = t_iP$, records the session identifier $\Pi_{i,j} = (I, ID_i, ID_j, (R_i, T_i), -)$, and returns $(R, ID_j, ID_i, (R_i, T_i))$.
2. *Send*$(R, ID_i, ID_j, (R_j, T_j))$: $\mathscr{C}$ checks if $T_j \in G^*$, then $\mathscr{C}$ chooses $t_i \in_R Z_q^*$, computes $T_i = t_iP$, records the session identifier $\Pi_{i,j} = (R, ID_i, ID_j, (R_i, T_i), (R_j, T_j))$, and returns the tuple $(I, ID_j, ID_i, (R_i, T_i))$. The session keys are computed as illustrated in Section 2.4.3, and the session is marked as completed.
3. *Send*$(I, ID_i, ID_j, (R_j, T_j))$: $\mathscr{C}$ checks if $T_j \in G^*$ and there exists an incomplete session $\Pi_{i,j} = (I, ID_i, ID_j, (R_i, T_i), -)$, then $\mathscr{C}$ completes the session by computing the session key and records the session identifier $\Pi_{i,j} = (I, ID_i, ID_j, (R_i, T_i), (R_j, T_j))$.

The oracle queries and secret reveal queries are simulated as follows.

1. $H_2(ID_i, ID_j, R_i, R_j, T_i, T_j, K_1, K_2, K_3)$: If the tuple $(ID_i, ID_j, R_i, R_j, T_i, T_j, K_1, K_2, K_3)$ is recorded in the $H_2$ list, then the corresponding value $SK$ is returned. Else, if the session $\Pi_{i,j} = (I, ID_i, ID_j, (R_i, T_i), (R_j, T_j))$ or the session $\Pi_{j,i} = (R, ID_j, ID_i, (R_j, T_j), (R_i, T_i))$ is recorded in the list $L_s$, then $\mathscr{C}$ checks the correctness of the shared secrets $(K_1, K_2, K_3)$ using the $DDH(*, *, *)$ oracle and the known private key value(s) and returns the session key $SK$, if the shared secrets are correctly formed. Else, $\mathscr{C}$ returns a random value $SK$ and records the same in the $L_{H_2}$ list.
2. *SessionKeyReveal*$(\Pi_{i,j})$: If the session $\Pi_{i,j}$ is not completed, $\mathscr{C}$ returns the error symbol. Else, if $\Pi_{i,j}$ is a completed session recorded in $L_s$, $\mathscr{C}$ returns the corresponding session key $SK$. Else, if the tuple $(ID_i, ID_j, R_i, R_j, T_i, T_j, K_1, K_2, K_3)$ is recorded in the $H_2$ list, $\mathscr{C}$ checks the correctness of the shared secrets $(K_1, K_2, K_3)$ using the $DDH(*, *, *)$ oracle and the known private key value(s) and returns the recorded session key value $SK$. Else, $\mathscr{C}$ returns a random value $SK$ and records the same in the $L_s$ list.
3. *EphemeralPrivateKeyReveal*$(\Pi_{i,j})$: If the ephemeral public key corresponding to $\Pi_{i,j}$ is $T_i = X$ or $T_i = Y$, where $(X, Y)$ is the CDH instance, then $\mathscr{C}$ aborts the simulation. Else, $\mathscr{C}$ returns the ephemeral private key $t_i \in Z_q^*$, corresponding to the session $\Pi_{i,j}$.
4. *StaticPrivateKeyReveal*$(ID_i)$: If the static public key $P_i = X$ or $P_i = Y$, where $(X, Y)$ is the CDH instance, then $\mathscr{C}$ aborts the simulation. Else, $\mathscr{C}$ returns the static private key $d_i \in Z_q^*$, corresponding to $ID_i$.
5. *CreatePeer*$(ID_i)$: $\mathscr{C}$ responds to the query faithfully, by generating the static public–private key pair of $ID_i$ and returning the same to $\mathscr{A}$. The entity $ID_i$ will be considered to be fully controlled by the adversary $\mathscr{A}$.
6. *Test*$(\Pi)$: For the event $W \wedge F_{1a}$, if $T_A \neq X$ or $P_B \neq Y$ for the test session $\Pi$, then $\mathscr{C}$ aborts the simulation. Otherwise, the simulation proceeds normally.

The probability that $\mathscr{C}$ chooses the session $\Pi$ with the initiating peer $u_A$, the responding peer $u_B$, the ephemeral public key $T_A = X$, and the static public key $P_B = Y$ is $\frac{1}{n^2m}$. The simulation remains perfect except with a negligible probability. $\mathscr{A}$ can query any static private key $d_i \neq d_B$, any ephemeral private key $t_i \neq t_A$ corresponding to $\Pi$, and $SK_{i,j} \neq SK_{A,B}$ corresponding to $\Pi$. If $\mathcal{A}$ wins the security game, then it must have issued the query $H_2(ID_A, ID_B, R_A, R_B, T_A, T_B, K_{AB_1}, K_{AB_2}, K_{AB_3})$. Using the known value $d_A$, $\mathcal{C}$ checks the correctness of the shared secrets $(K_{AB_1}, K_{AB_2}, K_{AB_3})$ as follows: $\mathcal{C}$ computes $K_1' = K_{AB_1} - 2d_A(T_B + 2P_B)$, $K_2' = K_{AB_2} + d_A(T_B - P_B)$, and $K_3' = K_{AB_3} - d_A(2T_B + P_B)$, and checks if $DDH(T_A, T_B + 2P_B, K_1') = 1$, $DDH(T_A, T_B - P_B, K_2') = 1$, and $DDH(2T_A, 2T_B + P_B, K_3') = 1$. If the validation succeeds, then $\mathcal{C}$ computes $CDH(X, Y) = \frac{(4K_1' - K_3')}{6}$. Thus the GDH solver succeeds with the probability $\Pr(\mathcal{C}) = \frac{P_1}{n^2 m h_2}$, where $P_1$ is the probability that the event $W \wedge F_{1a}$ occurs and $h_2$ denotes the number of hash queries to the oracle $H_2$.

### 2.5.2 EVENT $W \wedge F_{1B}$

In this case, the matching session $\Pi^*$ corresponding to the test session does not exist. $\mathcal{A}$ can issue the *EphemeralPrivateKeyReveal* ($\Pi$) query, but cannot reveal the static private keys of the participating peers $(u_A, u_B)$ corresponding to the test session $\Pi$. For any entity $u_i$, such that $u_i \neq u_A$ and $u_i \neq u_B$, $\mathcal{C}$ chooses random $(d_i, h_i) \in_R Z_q^*$ and sets $R_i = d_i P - h_i P_{Pub}$. Thus the static public key $P_i = d_i P = R_i + h_i P_{Pub}$, as per the protocol specification. For the entity $u_A(u_B)$, $\mathcal{C}$ chooses $h_A(h_B) \in_R Z_q^*$ and sets $R_A = X - h_A P_{pub}$ ($R_B = X - h_B P_{pub}$, respectively). Thus the static public key of $u_A$ is set as $P_A = X$, and the static public key of $u_B$ is set as $P_B = Y$. $\mathcal{C}$ responds to the oracle queries and the user activations in the same way as in the event $W \wedge F_{1b}$. The GDH solver chooses the session $\Pi$ with the initiating peer $u_A$ and the responding peer $u_B$, as the test session, with the probability $\frac{1}{n^2}$. The simulation remains perfect except with a negligible probability. $\mathcal{A}$ can query any static private key $d_i \neq d_A \wedge d_i \neq d_B$ and any session key $SK_{i,j} \neq SK_{A,B}$ corresponding to $\Pi$. If $\mathcal{A}$ wins the security game, then it must have issued the query $H_2(ID_A, ID_B, R_A, R_B, T_A, T_B, K_{AB_1}, K_{AB_2}, K_{AB_3})$. Using the known value $t_A$, $\mathcal{C}$ checks the correctness of the shared secrets $(K_{AB_1}, K_{AB_2}, K_{AB_3})$ as follows: $\mathcal{C}$ computes $K_1' = K_{AB_1} - t_A(T_B + 2P_B)$, $K_2' = t_A(T_B - P_B) - K_{AB_2}$, and $K_3' = K_{AB_3} - 2t_A(2T_B + P_B)$, and checks if $DDH(2P_A, T_B + 2P_B, K_1') = 1$, $DDH(P_A, T_B - P_B, K_2') = 1$, and $DDH(P_A, 2T_B + P_B, K_3') = 1$. If the validation succeeds, then $\mathcal{C}$ computes $CDH(X, Y) = \frac{(K_1' - 2K_2')}{6}$. Thus the GDH solver $\mathcal{C}$ succeeds with the probability $\Pr(\mathcal{C}) = \frac{P_2}{n^2 h_2}$, where $P_2$ is the probability that the event $W \wedge F_{1b}$ occurs and $h_2$ denotes the number of hash queries to the oracle $H_2$.

### 2.5.3 EVENT $W \wedge F_{2A}$

In this case, the matching session $\Pi^*$ exists; therefore $\mathcal{A}$ is a passive adversary that faithfully conveys the protocol messages. $\mathcal{A}$ cannot issue either *StaticPrivateKeyReveal*($ID_A$) or *StaticPrivateKeyReveal* ($ID_B$). The GDH solver $\mathcal{C}$ embeds the CDH instance $(X, Y)$ in the same way as in the event $W \wedge F_{1b}$. Therefore the static public key of $u_A$ is set as $P_A = X$, and the static public key of $u_B$ is set as $P_B = Y$. The user activations and oracle queries are also similar to those of the event $W \wedge F_{1b}$. However, the simulation is comparatively easier than that of the event $W \wedge F_{1b}$, since $\mathcal{C}$ knows both ephemeral private keys $(t_A, t_B)$. If $\mathcal{A}$ wins the security game, then it must have issued the query $H_2(ID_A, ID_B, R_A, R_B, T_A, T_B, K_{AB_1}, K_{AB_2}, K_{AB_3})$. Using the known values $(t_A, t_B)$, $\mathcal{C}$ checks the correctness of the shared secrets $(K_{AB_1}, K_{AB_2}, K_{AB_3})$ as follows: $\mathcal{C}$ computes $K_1' = K_{AB_1} - t_A(T_B + 2P_B) - 2t_B P_A$,

$K_2^{'} = t_A(T_B - P_B) - K_{AB_2} - t_B P_A$, and $K_3^{'} = K_{AB_3} - 2t_A(2T_B + P_B) - 2t_B P_A$, and checks if $DDH(2P_A, 2P_B, K_1^{'}) = 1$, $DDH(P_A, P_B, -K_2^{'}) = 1$, and $DDH(P_A, P_B, K_3^{'}) = 1$. If the validation succeeds, then $\mathscr{C}$ computes $CDH(X, Y) = K_3^{'}$. Thus the GDH solver $\mathscr{C}$ succeeds with the probability $\Pr(\mathscr{C}) = \frac{P_3}{n^2 h_2}$, where $P_3$ is the probability that the event $W \wedge F_{2a}$ occurs and $h_2$ denotes the number of hash queries to the oracle $H_2$.

## 2.5.4 EVENT $W \wedge F_{2B}$

In this case, the matching session $\Pi^*$ exists, and $\mathscr{A}$ cannot issue either *StaticPrivateKeyReveal*($ID_A$) or *EphemeralPrivateKeyReveal*($\Pi^*$). $\mathscr{C}$ chooses the $l^{th}$ session $\Pi$ between the initiating peer $u_A$ and the responding peer $u_B$, as the test session, with a probability $\frac{1}{n^2 m}$. C assigns static private keys to all users $u_i \neq u_A$, by choosing random $(d_i, h_i) \in_R Z_q^*$ and setting $R_i = d_i P - h_i P_{Pub}$. Thus the static public key $P_i = d_i P = R_i + h_i P_{Pub}$, as per the protocol specification. For the entity $u_A$, $\mathscr{C}$ chooses $h_A \in_R Z_q^*$ and sets $R_A = X - h_A P_{pub}$. $\mathscr{C}$ also sets the ephemeral public key of session $\Pi^*$ as $T_B = Y$. $\mathscr{C}$ responds to the oracle queries and the user activations in the same way as in the event $W \wedge F_{1b}$. If $\mathscr{A}$ wins the security game, then it must have issued the query $H_2(ID_A, ID_B, R_A, R_B, T_A, T_B, K_{AB_1}, K_{AB_2}, K_{AB_3})$. Using the known values $(t_A, d_B)$, $\mathscr{C}$ checks the correctness of the shared secrets $(K_{AB_1}, K_{AB_2}, K_{AB_3})$ as follows: $\mathscr{C}$ computes $K_1^{'} = K_{AB_1} - t_A(T_B + 2P_B) - 4d_B P_A$, $K_2^{'} = t_A(T_B - P_B) - K_{AB_2} + d_B P_A$, and $K_3^{'} = K_{AB_3} - 2t_A(2T_B + P_B) - d_B P_A$, and checks if $DDH(2P_A, T_B, K_1^{'}) = 1$, $DDH(P_A, T_B, K_2^{'}) = 1$, and $DDH(P_A, 2T_B, K_3^{'}) = 1$. If the validation succeeds, then $\mathscr{C}$ computes $CDH(X, Y) = K_2^{'}$. Thus the GDH solver $\mathscr{C}$ succeeds with the probability $\Pr(\mathscr{C}) = \frac{P_4}{n^2 m h_2}$, where $P_4$ is the probability that the event $W \wedge F_{2b}$ occurs and $h_2$ denotes the number of hash queries to the oracle $H_2$.

## 2.5.5 EVENT $W \wedge F_{2C}$

In this case, the matching session $\Pi^*$ exists, and $\mathscr{A}$ cannot issue either *EphemeralPrivateKeyReveal*($\Pi$) or *StaticPrivateKeyReveal*($ID_B$). $\mathscr{C}$ chooses the $l^{th}$ session $\Pi$ between the initiating peer $u_A$ and the responding peer $u_B$, as the test session, with a probability $\frac{1}{n^2 m}$. The simulation closely resembles that of the event $W \wedge F_{2b}$, except that the roles of $u_A$ and $u_B$ are interchanged. Thus, for the entity $u_B$, $\mathscr{C}$ chooses $h_B \in_R Z_q^*$ and sets $R_B = Y - h_B P_{pub}$. $\mathscr{C}$ also sets the ephemeral public key of session $\Pi$ as $T_A = X$. $\mathscr{C}$ responds to the oracle queries and the user activations in the same way as in the event $W \wedge F_{1a}$. If $\mathscr{A}$ wins the security game, then it must have issued the query $H_2(ID_A, ID_B, R_A, R_B, T_A, T_B, K_{AB_1}, K_{AB_2}, K_{AB_3})$. Using the known values $(t_B, d_A)$, $\mathscr{C}$ checks the correctness of the shared secrets $(K_{AB_1}, K_{AB_2}, K_{AB_3})$ as follows: $\mathscr{C}$ computes $K_1^{'} = K_{AB_1} - t_B(T_A + 2P_A) - 2d_A P_B$, $K_2^{'} = K_{AB_2} - t_B(T_A - P_A) - d_A P_B$, and $K_3^{'} = K_{AB_3} - 2t_B (2T_A + P_A) - d_A P_B$, and checks if $DDH(T_A, 2P_B, K_1^{'}) = 1$, $DDH(T_A, P_B, -K_2^{'}) = 1$, and $DDH(2T_A, P_B, K_3^{'}) = 1$. If the validation succeeds, then $\mathscr{C}$ computes $CDH(X, Y) = -K_2^{'}$. Thus the GDH solver $\mathscr{C}$ succeeds with the probability $\Pr(\mathscr{C}) = \frac{P_5}{n^2 m h_2}$, where $P_5$ is the probability that the event $W \wedge F_{2c}$ occurs and $h_2$ denotes the number of hash queries to the oracle $H_2$.

## 2.5.6 EVENT $W \wedge F_{2D}$

In this case, the matching session $\Pi^*$ exists, and $\mathscr{A}$ cannot issue either *EphemeralPrivateKeyReveal*($\Pi$) or *EphemeralPrivateKeyReveal*($\Pi^*$). $\mathscr{C}$ assigns static private keys to each user $u_i$, by choosing random $(d_i, h_i) \in_R Z_q^*$ and setting $R_i = d_iP - h_iP_{Pub}$. Thus the static public key $P_i = d_iP = R_i + h_iP_{Pub}$, as per the protocol specification. $\mathscr{C}$ sets the ephemeral public key of the session $\Pi$ as $T_A = X$ and the ephemeral private key of the session $\Pi^*$ as $T_B = Y$. $\mathscr{C}$ responds to the oracle queries and the user activations as illustrated in the event $W \wedge F_{1a}$. $\mathscr{C}$ chooses the $l$th session $\Pi$ between the initiating peer $u_A$ and the responding peer $u_B$, as the test session and $\Pi^*$ as the matching session, with a probability $\frac{1}{n^2m^2}$. If $\mathscr{A}$ wins the security game, then it must have issued the query $H_2(ID_A, ID_B, R_A, R_B, T_A, T_B, K_{AB_1}, K_{AB_2}, K_{AB_3})$. Using the known values $(d_A, d_B)$, $\mathscr{C}$ checks the correctness of the shared secrets $(K_{AB_1}, K_{AB_2}, K_{AB_3})$ as follows: $\mathscr{C}$ computes $K_1^{'} = K_{AB_1} - 2d_A(T_B + 2P_B) - 2d_BT_A$, $K_2^{'} = K_{AB_2} + d_A(T_B - P_B) + d_BT_A$, and $K_3^{'} = K_{AB_3} - d_A(2T_B + P_B) - 2d_BT_A$, and checks if $DDH(T_A, T_B, K_1^{'}) = 1$, $DDH(T_A, T_B, K_2^{'}) = 1$, and $DDH(2T_A, 2T_B, K_3^{'}) = 1$. If the validation succeeds, then $\mathscr{C}$ computes $CDH(X, Y) = K_1^{'}$. Thus the GDH solver $\mathscr{C}$ succeeds with the probability $\Pr(\mathscr{C}) = \frac{P_6}{n^2mh_2}$, where $P_6$ is the probability that the event $W \wedge F_{2d}$ occurs and $h_2$ denotes the number of hash queries to the oracle $H_2$.

Therefore, if $\mathscr{A}$ succeeds with a non-negligible probability in any of the complementary cases discussed above, then the GDH solver $\mathscr{C}$ also succeeds in breaking the GDH assumption with a nonnegligible probability. This contradicts the GDH assumption that no PPT algorithm can solve the GDH problem except with a nonnegligible probability. Theorem 1 is hence proved.

## 2.6 ANALYSIS OF DANG ET AL.'S IDENTITY-BASED AUTHENTICATED KEY AGREEMENT PROTOCOL

The protocol proposed by DXC [13] is summarized below. Initially, during the setup phase, the KGC defines the hash functions: $H_1{:}\{0,1\}^* \times G \rightarrow Z_q^*$ and $H_2{:}\{0,1\}^* \times \{0,1\}^* \times G^5 \rightarrow \{0,1\}^z$, where $z$ denotes the size of the final session key. The setup and entity registration phase is similar to that of the proposed scheme. The key agreement phase between the vehicles $u_A$ and $u_B$, with the identities $ID_A$ and $ID_B$, proceeds as follows.

- Exchange of session-specific ephemeral keys:
  - Entity $u_A$ chooses $t_A \in_R Z_q^*$ as the ephemeral private key and computes the ephemeral public keys $T_A{}^1 = t_AR_A$ and $T_A{}^2 = t_AP_{pub}$. The tuple $(ID_A, R_A, T_A{}^1, T_A{}^2)$ is sent to the entity $u_B$.
  - On receiving the tuple $(ID_A, R_A, T_A{}^1, T_A{}^2)$, the entity $u_B$ chooses $t_B \in_R Z_q^*$ as the ephemeral private key and computes the ephemeral public keys $T_B{}^1 = t_BR_B$ and $T_B{}^2 = t_BP_{pub}$. The tuple $(ID_B, R_B, T_B{}^1, T_B{}^2)$ is sent to the entity $u_A$.
- Session key establishment:
  - Entity $u_A$ computes the shared secret $K_{AB} = t_Ad_A(T_B{}^1 + H_1(ID_B, R_B)T_B{}^2)$ and session key $SK_{AB} = H_2(ID_A, ID_B, T_A{}^1, T_A{}^2, T_B{}^1, T_B{}^2, K_{AB})$.
  - Entity $u_B$ computes the shared secret $K_{BA} = t_Bd_B(T_A{}^1 + H_1(ID_A, R_A)T_A{}^2)$ and session key $SK_{BA} = H_2(ID_A, ID_B, T_A{}^1, T_A{}^2, T_B{}^1, T_B{}^2, K_{BA})$.

The correctness of the protocol can be validated as follows: $K_{AB} = t_Ad_A\left(T_B{}^1 + H_1(ID_B, R_B)T_B{}^2\right) = t_Ad_At_B\left(R_B + H_1(ID_B, R_B)P_{pub}\right) = t_Ad_At_Bd_BP = t_Bd_B(t_AR_A + t_AH_1(ID_A, R_A)P_{pub}) = t_Bd_B(T_A{}^1 + H_1(ID_A, R_A)T_A{}^2) = K_{BA}$.

### 2.6.1 KEY COMPROMISE IMPERSONATION ATTACK AGAINST DANG ET AL.'S PROTOCOL

Consider that the adversary obtains the static private key $d_B$ of the vehicle $u_B$. The adversary impersonates another entity $u_A$, to the compromised entity $u_B$ as follows.

- Exchange of session-specific ephemeral keys:
  - Initially, the adversary generates the ephemeral public keys $\overline{T}_A^{\,1} = P_{pub}H_1(ID_A, R_A) + d_B^{-1}P$ and $\overline{T}_A^{\,2} = -P_{pub}$. This computation is possible, since $(ID_A, R_A)$ can be obtained by observing any of the previous communications by the entity $I$, and $P_{pub}$ is the widely distributed master public key. The adversary then forwards the tuple $(ID_A, R_A, \overline{T}_A^{\,1}, \overline{T}_A^{\,2})$ to the entity $u_B$.
  - On receiving $(ID_A, R_A, \overline{T}_A^{\,1}, \overline{T}_A^{\,2})$, the entity $u_B$ executes the protocol in the conventional manner. Entity $u_B$ chooses $t_B \in_R Z_q^*$ and generates the ephemeral public keys $T_B^{\,1} = t_B R_B$ and $T_B^{\,2} = t_B P_{pub}$. The tuple $(ID_B, R_B, T_B^{\,1}, T_B^{\,2})$ is sent back to the initiator.
- Session key establishment:
  - The adversary computes the shared secret as $K_{AB} = d_B^{-1}\left(T_B^{\,1} + H_1(ID_B, R_B)T_B^{\,2}\right) = d_B^{-1}t_B\left(R_B + H_1(ID_B, R_B)P_{pub}\right) = d_B^{-1}t_B d_B P = t_B P$. The session key is computed as $SK_{AB} = H_2(ID_A, ID_B, \overline{T}_A^{\,1}, \overline{T}_A^{\,2}, T_B^{\,1}, T_B^{\,2}, K_{AB})$.
  - The victim $u_B$ computes the shared secret $K_{BA} = t_B d_B\left(\overline{T}_A^{\,1} + H_1(ID_A, R_A)\overline{T}_A^{\,2}\right) = t_B d_B\left(P_{pub}H_1(ID_A, R_A) + d_B^{-1}\right.$ $\left.P - H_1(ID_A, R_A)P_{pub}\right) = t_B d_B d_B^{-1}P = t_B P = K_{AB}$. The session key is computed as $SK_{BA} = SK_{AB} = H_2(ID_A, ID_B, \overline{T}_A^{\,1}, \overline{T}_A^{\,2}, T_B^{\,1}, T_B^{\,2}, K_{AB})$.

It is apparent that both the shared secrets and session keys are equivalent. Thus the adversary succeeds in the KCI attack against the compromised entity $u_B$. Next, we illustrate, with examples, how a KCI attack can be exploited to perpetrate fatal road traffic collisions.

As shown in Fig. 2.3, in the first case, the legitimate vehicle A detects the ongoing manhole maintenance work and sends the "Lane closed" alert message to the trailing vehicle B. However, a malicious entity E that possesses B's static private key impersonates A to B (KCI attack) and alters the message as "Lane clear." This can apparently jeopardize the vehicle B. The second example illustrates a truck platooning system wherein the driverless automated trailing trucks B and C are operated by the controlling truck A. In this case, an adversary E that possesses the static private key of the truck B can wreak havoc, by impersonating the controller A to the compromised entity B (KCI attack) and simultaneously impersonating B to C. The latter attack is trivial, since E already knows B's private key and can impersonate B to any other entity, including C. Thus B is coerced into thinking that A is halting and triggers its emergency braking system, while C is instructed to move forward at the same pace, ultimately leading to traffic collisions.

### 2.6.2 FLAWS IN THE SECURITY PROOF

For cases $S_5$ and $S_6$ in the security proof of [13], the simulator is dealing with an active adversary that chooses the ephemeral private key $t_B$, corresponding to the session responder $ID_B$. Hence, the simulator is unaware of the value $t_B$.

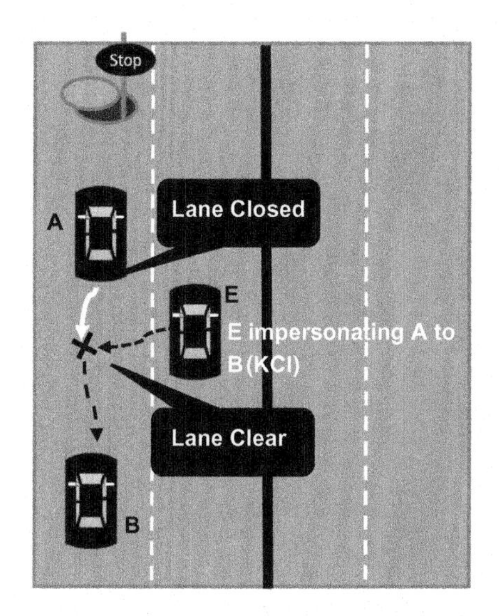


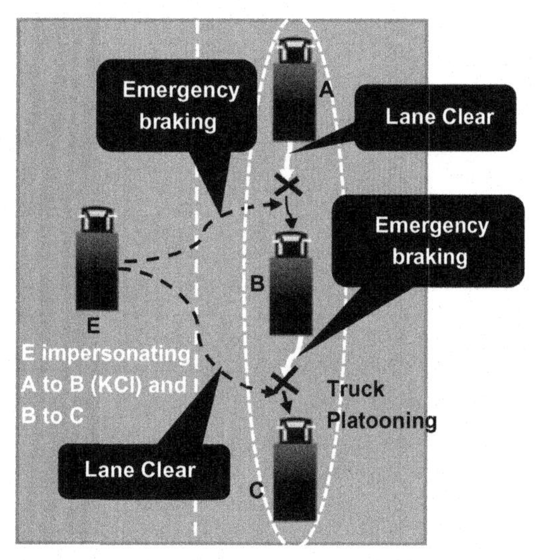


**FIGURE 2.3**

Road traffic scenarios illustrating successful key compromise impersonation attacks that can lead to fatalities.

•*Case $S_5$: The adversary does not query $ID_A$'s ephemeral private key or $ID_B$'s static private key. Matching session does not exist.* In this case, according to the proof, the simulator embeds the CDH instance $(X, Y)$ as $P_{pub} = X$ and $T_A = Y$. The proof assumes that the incoming tuple from the adversary is of the form $\left(ID_B, R_B', T_B{}^1 = (t_B P_B - t_B H_1(ID_B, R_B')X), T_B{}^2 = t_B X\right)$, where $R_B'$ might be the incorrect public key material of $u_B$. However, the simulator cannot expect the adversary to send this tuple to favor the simulation, since the adversary might follow the protocol specification to send the tuple $\left(ID_B, R_B', T_B{}^1 = t_B R_B', T_B{}^2 = t_B X\right)$. Moreover, the simulator cannot solve for the GDH instance, as the values $(t_A, t_B, d_B)$ are unknown. Specifically, let $H_1(ID_B, R_B') = h$ and $K_{AB} = DLOG(Y) d_A (t_B R_B' + t_B X h)$. After the oracle replay, let $H_1(ID_B, R_B') = \overline{h}$ and $\overline{K_{AB}} = DLOG(Y) d_A (t_B R_B' + t_B X \overline{h})$. Then, $K_{AB} - \overline{K_{AB}} = DLOG(Y) d_A t_B X(h - \overline{h})$. The simulator cannot compute $GDH(X, Y)$, unless it knows the value of $t_B$ chosen by the adversary.

•*Case $S_6$: The adversary does not query $ID_A$'s static private key or $ID_B$'s static private key. Matching session does not exist.* The simulator embeds the CDH instance $(X, Y)$ as $P_{pub} = X$ and $P_A = Y$. The proof assumes that the incoming tuple from the adversary is of the form $\left(ID_B, R_B', T_B{}^1 = (t_B P_B - t_B H_1(ID_B, R_B')X), T_B{}^2 = t_B X\right)$, where $R_B'$ might be the incorrect public key material. As in the previous case, this assumption is incorrect, and the simulator fails to compute the GDH instance, since the values $(d_A, d_B, t_B)$ are unknown.

## 2.7 EFFICIENCY ANALYSIS

The proposed ID-AK A scheme for V2V communications is highly efficient, since it involves only elliptic curve scalar multiplications and point additions. Table 2.1 presents a comparative analysis

**Table 2.1 Comparison of the security properties, and communication and computational overheads of the proposed scheme with the existing ID-AKA protocols.**

| Protocol | Assum. | Model | wPFS | $R_{KCI}$ | $R_{ESKL}$ | Message Size $u_A \rightarrow u_B, u_A \leftarrow u_B$ | Group Operations | Comm. Cost (Bytes) | Comp. Cost (ms) |
|---|---|---|---|---|---|---|---|---|---|
| FG [22] | SDH | CK | Yes | Yes | No | $2G, 2G$ | $4PM + 2PA$ | (128, 128) | 13.89 |
| CKD [23] | GDH | mBR | Yes | Yes | No | $2G, 2G$ | $5PM + 2PA$ | (128, 128) | 17.33 |
| $IB_1$ [24] | ≈ | ≈ | ≈ | ≈ | ≈ | $2G, (2G, \{0,1\}^z)$ | $4PM + 2PA$ | (128, 160) | 13.89 |
| XW [25] | CDH | CK | Yes | ≈ | No | $(3G, 1Z_q^*),(3G, 1Z_q^*)$ | $6PM + 3PA$ | (212, 212) | 20.82 |
| SWZ [26] | GDH | eCK | Yes | Yes | Yes | $2G, 2G$ | $6PM + 3PA$ | (128, 128) | 20.82 |
| BSV [27] | GDH | $eCK^!$ | Yes | Yes | No | $2G, 2G$ | $4PM + 2PA$ | (128, 128) | 13.89 |
| $NCL_1$ [28] | CDH | eCK | Yes | Yes | Yes | $3G, 3G$ | $7PM + 6PA$ | (192, 192) | 24.30 |
| $NCL_2$ [28] | GDH | eCK | Yes | Yes | Yes | $3G, 3G$ | $5PM + 4PA$ | (192, 192) | 17.35 |
| $IB_2$ [29] | ≈ | BAN | No | ≈ | ≈ | $(2G, 1Z_q^*), (2G, 1Z_q^*)$ | $5PM + 2PA$ | (148, 148) | 17.33 |
| DXC [13] | GDH | $eCK^!$ | Yes | No | Yes | $3G, 3G$ | $4PM + 1PA$ | (192, 192) | 13.88 |
| Proposed | GDH | eCK | Yes | Yes | Yes | $2G, 2G$ | $5PM + 6PA$ | (128, 128) | 17.38 |

*$R_{ESKL}$, Resistance to ESKL attack; $R_{KCI}$, resistance to KCI attack;* ≈, *lack of provable security;* Assum., *hardness assumption, model—security model; BSV, Bala et al.; CDH, computational Diffie–Hellman; CK, Canetti and Krawczyk; CKD, Cao et al.;* Comm., *cost–communication overhead in bytes;* Comp., *cost–computation overhead in milliseconds; DXC, Dang et al.;* $eCK^!$, *incorrect security proof; eCK, extended CK; FG, Fiore and Gennaro; SDH, Strong Diffie Hellman; GDH, gap Diffie–Hellman; $IB_2$, Islam and Biswas;* ID-AKA, *identity-based authenticated key agreement; NCL, Ni et al.;* PA, *point addition;* PM, *point multiplication;* wPFS, *weak perfect forward secrecy; XW, Xie and Wang.*

of the security properties (security model, hardness assumption, compliance to wPFS property, and resilience to KCI as well as ESKL attacks) and performance (communication and computational overheads) of the proposed scheme, with the existing ID-AKA protocols in the literature.

Specifically, the bilinear pairing-free ID-AKA protocol proposed by Fiore and Gennaro (FG) [22] is proven secure in the CK security model and lacks ESKL resilience [30]. Cao et al. (CKD) [23] proposed an ID-AKA scheme based on the GDH assumption in the weaker mBR model. Islam and Biswas ($IB_1$) proved that CKD is susceptible to ESKL attacks and proposed an alternative scheme with ESKL resilience [24]. However, $IB_1$ lacks provable security. Xie and Wang [25] (XW) proposed a pairing-free ID-AKA scheme with wPFS property; however, the protocol requires signature generation and verification by each participating peer to ensure security. Furthermore, Ni et al. [28] noted that the protocol is also susceptible to ESKL attacks illustrated in [30]. Bala et al. [27] (BSV) proposed an efficient ID-AKA protocol that is similar to FG; hence, the protocol is also vulnerable to ESKL attacks. The pairing-free ID-AKA scheme by Islam and Biswas ($IB_2$) [29] is analyzed by using the Burrows, Abadi and Needham (BAN) logic; hence, it does not ensure provable security against KCI and ESKL attacks or compliance to wPFS property. As shown in Section 2.6, the ID-AKA scheme proposed by (DXC) [13], for securing V2V communications, is vulnerable to KCI attacks. Few of these protocols are computationally more efficient than the proposed scheme [13,22,24,27,29]. Nevertheless, none of these protocols can be deployed for securing the safety-critical communications in V2V ad hoc networks, since they exhibit conspicuous security vulnerabilities.

Since the computational power of nodes in a VANET is unconstrained, the execution time of each protocol in milliseconds is determined by using the pairing based cryptography (PBC) library, in a 64-bit Intel core-i5 machine, with a clock speed of 2.20 GHz, 8-GB RAM, and Windows 10 operating system. The elliptic curve group $G$ is defined over a supersingular curve (type A), with a curve equation $y^2 = x^3 + x$, embedding degree 2 and prime order $q = 160$ bits, at 80-bit security level. The size of each group element in $G$ is 512 bits, and the size of the final session key $z = 256$ bits.

As for the existing eCK secure protocols [26,28], the computational cost of the Sun, Wen and Zhang (SWZ) scheme [26] is much higher than the proposed scheme. Furthermore, the provably secure ID-AKA protocol proposed by Ni et al. ($NCL_1$ and $NCL_2$) [28] has higher communication and computational overhead than the proposed protocol. Although the computational cost of $NCL_2$ is comparable with that of the proposed scheme, the communication cost is higher by 64 bytes. Considering the highly dynamic nature of V2V communications, it is essential to minimize the message size to expedite the packet delivery rate. Hence, when compared with the existing ID-AKA schemes in the literature, the proposed protocol is highly suitable for securing intervehicular communications, as it maintains a fine balance between security and efficiency.

## 2.8 CONCLUSION

In this chapter, we cryptanalyze the recently proposed ID-AKA scheme (DXC) [13] for securing V2V communications. We prove that the protocol is vulnerable to KCI attacks and illustrate, with real-time examples, how an active adversary can benefit from this security vulnerability, to create road traffic collisions. The flaws in the security proof of [13] are also analyzed. Considering the highly dynamic topology of V2V networks, the brevity of the established sessions, and the momentousness of the exchanged automotive information, it is crucial to design a provably secure ID-AKA protocol for securing intervehicular communications. To this end, we propose an eCK-secure ID-AKA protocol with wPFS property and provable security against BI, KCI, UKS, and ESKL attacks. The proposed scheme requires only five scalar multiplications per-party for session key establishment and incurs minimal communication overheads. Thus, compared with the existing ID-AKA protocols in the literature, the proposed scheme achieves a fine tradeoff between security and efficiency.

## ACKNOWLEDGMENT

This work was funded by the Ministry of Electronics and Information Technology, Government of India, through the Visvesvaraya Ph.D. scheme for Electronics and IT.

## REFERENCES

[1] A. Wasef, Y. Jiang, X. Shen, DCS: an efficient distributed-certificate-service scheme for vehicular networks, IEEE Trans. Veh. Technol. 59 (2010) 533–549. Available from: https://doi.org/10.1109/TVT.2009.2028893.

[2] J. Huang, L. Yeh, C. Hung-Yu, ABAKA : an anonymous batch authenticated and key agreement scheme for value-added services in vehicular ad hoc networks, IEEE Trans. Veh. Technol. 60 (2011) 248–262. Available from: https://doi.org/10.1016/j.proeng.2012.01.204.

[3] C. Zhang, X. Lin, R. Lu, P.-H. Ho, X. Shen, An efficient message authentication scheme for vehicular communications, IEEE Trans. Veh. Technol. 57 (2008) 3357–3368. Available from: https://doi.org/10.1109/TVT.2008.928581.

[4] C. Buttner, S.A. Huss, A novel anonymous authenticated key agreement protocol for vehicular ad hoc networks, in: Proceedings of the 1st International Conference on Information Systems Security and Privacy (ICISSP 2015), 2015 <https://www.engineeringvillage.com/share/document.url?mid = inspec_3bbed980150d4810a65M775710178163171&database = ins>.

[5] C. Büttner, S.A. Huss, Real-world evaluation of an anonymous authenticated key agreement protocol for vehicular ad-hoc networks, in: IEEE 11th International Conference on Wireless and Mobile Computing, Networking and Communications. Real-World, 2015, pp. 259–269 <http://dblp.uni-trier.de/db/conf/icissp/icissp2015.html#ButtnerH15>.

[6] C.T. Li, M.S. Hwang, Y.P. Chu, A secure and efficient communication scheme with authenticated key establishment and privacy preserving for vehicular ad hoc networks, Comput. Commun. 31 (2008) 2803–2814. Available from: https://doi.org/10.1016/j.comcom.2007.12.005.

[7] M.J. Hinek, On the security of multi-prime RSA, J. Math. Cryptol. 2 (2008) 117–147. Available from: https://doi.org/10.1016/S0019-9958(82)90401-6.

[8] M. Ciet, F. Koeune, F. Laguillaumie, J.-J. Quisquater, Short private exponent attacks on fast variants of RSA, UCL Crypto Group Technical Report Series CG-2002/4, University Catholique de Louvain, 2002, pp. 1–24.

[9] R. Lu, X. Lin, H. Zhu, P.H. Ho, X. Shen, ECPP: efficient conditional privacy preservation protocol for secure vehicular communications, in: Proceedings of the IEEE INFOCOM, Phoenix, Arizona, USA, 2008, pp. 1903–1911. doi:10.1109/INFOCOM.2007.179.

[10] Y. Sun, R. Lu, X. Lin, X. Shen, J. Su, An efficient pseudonymous generation scheme with privacy preservation for vehicular communication, IEEE Trans. Veh. Technol. 59 (2010) 3589–3603. Available from: https://doi.org/10.1109/ICICA.2014.32.

[11] C. Chen, T. Hsu, H. Wu, J.Y. Chiang, W. Hsieh, Anonymous authentication and key-agreement schemes in vehicular ad-hoc networks, J. Internet Technol. 15 (2014) 893–902. Available from: https://doi.org/10.6138/JIT.2014.15.6.02.

[12] Y. Liu, Y. Wang, G. Chang, Efficient privacy-preserving dual authentication and key agreement scheme for secure V2V communications in an IoV paradigm, IEEE Trans. Intell. Transp. Syst. 18 (2017) 2740–2749.

[13] L. Dang, J. Xu, X. Cao, H. Li, J. Chen, Y. Zhang, Efficient identity-based authenticated key agreement protocol with provable security for vehicular ad hoc networks, Int. J. Distrib. Sens. Netw. 14 (2018). Available from: https://doi.org/10.1177/1550147718772545.

[14] B. Lamacchia, K. Lauter, A. Mityagin, Stronger security of authenticated key exchange, in: International Conference on Provable Security, Springer, Berlin, Heidelberg, 2007, pp. 1–16.

[15] M. Bellare, P. Rogaway, Entity authentication and key distribution, Annual International Cryptology Conference, Springer, Berlin, Heidelberg, 1993, pp. 232–249.

[16] M. Bellare, P. Rogaway, Provably secure session key distribution: the three party case, Proceedings of the Twenty-Seventh Annual ACM Symposium Theory of Computing, ACM, 1995, pp. 57–66.

[17] S. Blake-wilson, D. Johnson, A. Menezes, Key agreement protocols and their security analysis, IMA International Conference on Cryptography and Coding, Springer, Berlin, Heidelberg, 1997, pp. 30–45.

[18] R. Canetti, H. Krawczyk, Analysis of key-exchange protocols and their use for building secure channels, International Conference on the Theory and Applications of Cryptographic Techniques, Springer, Berlin, Heidelberg, 2001, pp. 453–474.

[19] C.J.F. Cremers, Formally and practically relating the CK, CK-HMQV, and eCK security models for authenticated key exchange, IACR Cryptol. EPrint Arch. 253 (2009) 1–19.

[20] B. Lamacchia, K. Lauter, A. Mityagin, Stronger security of authenticated key exchange, in: International Conference on Provable Security 2007, Wollongong, NSW, Australia , LNCS, Vol. 4784, Springer-Verlag, 2007, pp. 1–16. doi:10.1007/978-3-540-75670-5-1.

[21] C.P. Schnorr, Efficient identification and signatures for smart cardsLNCS 435 Advances in Cryptology-CRYPT0 ‘89, Santa Barbara, California, USA, Springer, 1990pp. 239–252.

[22] D. Fiore, R. Gennaro, Making the Diffie–Hellman protocol identity-based, in: Topics in Cryptology-CT-RSA 2010. CT-RSA 2010. Lecture Notes in Computer Science, vol. 5985, Springer, Berlin, Heidelberg, 2010, pp. 165–178. doi:10.1007/978-3-642-11925-5_12.

[23] X. Cao, W. Kou, X. Du, A pairing-free identity-based authenticated key agreement protocol with minimal message exchanges, Inf. Sci. 180 (2010) 2895–2903. Available from: https://doi.org/10.1016/j.ins.2010.04.002.

[24] S.K.H. Islam, G.P. Biswas, An improved pairing-free identity-based authenticated key agreement protocol based on ECC, Procedia Eng. 30 (2011) 499–507. Available from: https://doi.org/10.1016/j.proeng.2012.01.890.

[25] M. Xie, L. Wang, One-round identity-based key exchange with perfect forward secrecy, Inf. Process. Lett. 112 (2012) 587–591. Available from: https://doi.org/10.1016/j.ipl.2012.05.001.

[26] H. Sun, Q. Wen, H. Zhang, Z. Jin, A strongly secure identity-based authenticated key agreement protocol without pairings under the GDH assumption, Secur. Commun. Netw. 8 (2015) 3167–3179. Available from: https://doi.org/10.1002/sec.

[27] S. Bala, G. Sharma, A.K. Verma, PF-ID-2PAKA: pairing free identity-based two-party authenticated key agreement protocol for wireless sensor networks, Wirel. Pers. Commun. 87 (2016) 995–1012. Available from: https://doi.org/10.1007/s11277-015-2626-5.

[28] L. Ni, G. Chen, J. Li, Y. Hao, Strongly secure identity-based authenticated key agreement protocols without bilinear pairings, Inf. Sci. (Ny.) 367–368 (2016) 176–193. Available from: https://doi.org/10.1016/j.ins.2016.05.015.

[29] S.H. Islam, G.P. Biswas, A pairing-free identity-based two-party authenticated key agreement protocol for secure and efficient communication, J. King Saud. Univ. Comput. Inf. Sci. 29 (2017) 63–73. Available from: https://doi.org/10.1016/j.jksuci.2015.01.004.

[30] Q. Cheng, C. Ma, Ephemeral key compromise attack on the IB-KA protocol, IACR Cryptol. EPrint Arch. 568 (2009) 3–6.

CHAPTER

# 3 DYNAMIC SELF-AWARE TASK ASSIGNMENT ALGORITHM FOR AN INTERNET OF THINGS-BASED WIRELESS SURVEILLANCE SYSTEM

**Titus Issac[1], Salaja Silas[1], Elijah Blessing Rajsingh[1] and Sharmila Anand John Francis[2]**

*[1]Karunya Institute of Technology and Sciences, Coimbatore, India [2]King Khalid University, Abha, Saudi Arabia*

**CHAPTER OUTLINE**



## 3.1 INTRODUCTION

Surveillance is the process of monitoring and gathering vital information from an environment through keen observation. Surveillance has always been an integral part of the defence sector. Highly skilled people were recruited and deployed for the purpose of surveillance. However, the effectiveness of a human-based surveillance system has always been constrained by time, climatic conditions, etc.

The dawn of the information age led to the development of video-based surveillance by employing a set of sophisticated surveillance video cameras. The traditional video surveillance system, such as the CCTV, comprises of a set of video cameras to gather intelligence through the audio and video streams from the desired monitoring environment [1,2]. The video surveillance system

The Cognitive Approach in Cloud Computing and Internet of Things Technologies for Surveillance Tracking Systems.
DOI: https://doi.org/10.1016/B978-0-12-816385-6.00003-9

outperformed its human counterpart, as the efficiency of the video cameras was not constrained by environmental factors. But the analysis of the captured video and audio streams in the traditional video surveillance is cumbersome, as the earlier analog version of the closed circuit television (CCTV) system had no physical storages and every camera required a dedicated monitor [3].

Due to the recent digitization and video multiplexing techniques, the contemporary, smart surveillance system was developed. It utilized smart video cameras with onboard sophisticated sensors as well as a dedicated set of sensors, and digital video recorders (DVRs). The sensors in the smart surveillance system aided in capturing the environment on an on-demand basis and thereby reducing human intervention during the data analysis process [4]. The generated multimedia data were stored in DVR, enabling the analysis to be carried out in the desired time. Earlier, the video surveillance systems were confined to core military applications, but after the inception of the information age, it was extended for civilian use in their workplaces, homes, and vehicles [3].

The traditional video surveillance systems were based on wired systems. Considering the deployment, installation, and maintenance cost, wireless surveillance system (WSS) is preferred over the conventional wired surveillance system [5]. The core components in a WSS are as follows: (1) wireless video cameras (WVCs); (2) sensors; and (3) data centers. An example of an IoT-based smart multilayer wireless surveillance framework is depicted in Fig. 3.1, where a set of WVCs is connected to an intermediate data center called the mobile edge computing server (MECS). A mobile edge computing (MEC) cell is the maximum region within which a set of WVCs is able to connect to a MECS. A WSS may have one or more MEC cells based on the requirement of the application. The base layer comprises a set of WVCs and sensors, while the upper layer comprises of data centers with high power computing and storage.

A device with video recording capability is a primary part of the surveillance system. The block diagram of the various units in a WVC is depicted in Fig. 3.2. The storage unit of the WVC has inbuilt local storage and external storage capabilities. The local storage acts as temporary storage and enables buffering until the data are completely transmitted via the transmission media through the communication unit. The limited inbuilt storage of a WVC could be expanded as per the application requirements.

A WVC with higher processing capability has the inherent ability to preprocess and analyze the captured video. A WVC may also have a set of sensors and illumination units. The sensors such as presence sensor, luminous sensor, and infrared (IR) sensor are also embedded in the WVC [6]. Under lower lighting conditions, the onboard sensors trigger the IR light-emitting diodes to illuminate and capture an enhanced video. A WVC could be mounted with a wide range of wireless communication devices such as Zig-Bee, Wi-Fi, and 4G [7]. A battery-powered smart WVC is a highly energy-constrained device.

Apart from the embedded sensors of a WVC, a set of dedicated sensors is deployed optionally in preidentified critical points based on the requirement of a WSS. Widely used sensors in a WSS are temperature, humidity, acceleration, pressure, luminosity, motion, audio, and radiation [1]. Based on the activity, the sensors trigger the intermediate devices for further processing. The intermediate and top layers contain sophisticated servers that act as a storage repository, data processing unit, and prediction unit [1]. The major challenges of the WSS are (1) video size; (2) bandwidth; (3) computing; and (4) storage.

The Internet of Things (IoT) has revolutionized by linking digital and physical entities together [8]. The "thing" in the IoT represents a device connected to the Internet. In recent years, the IoT

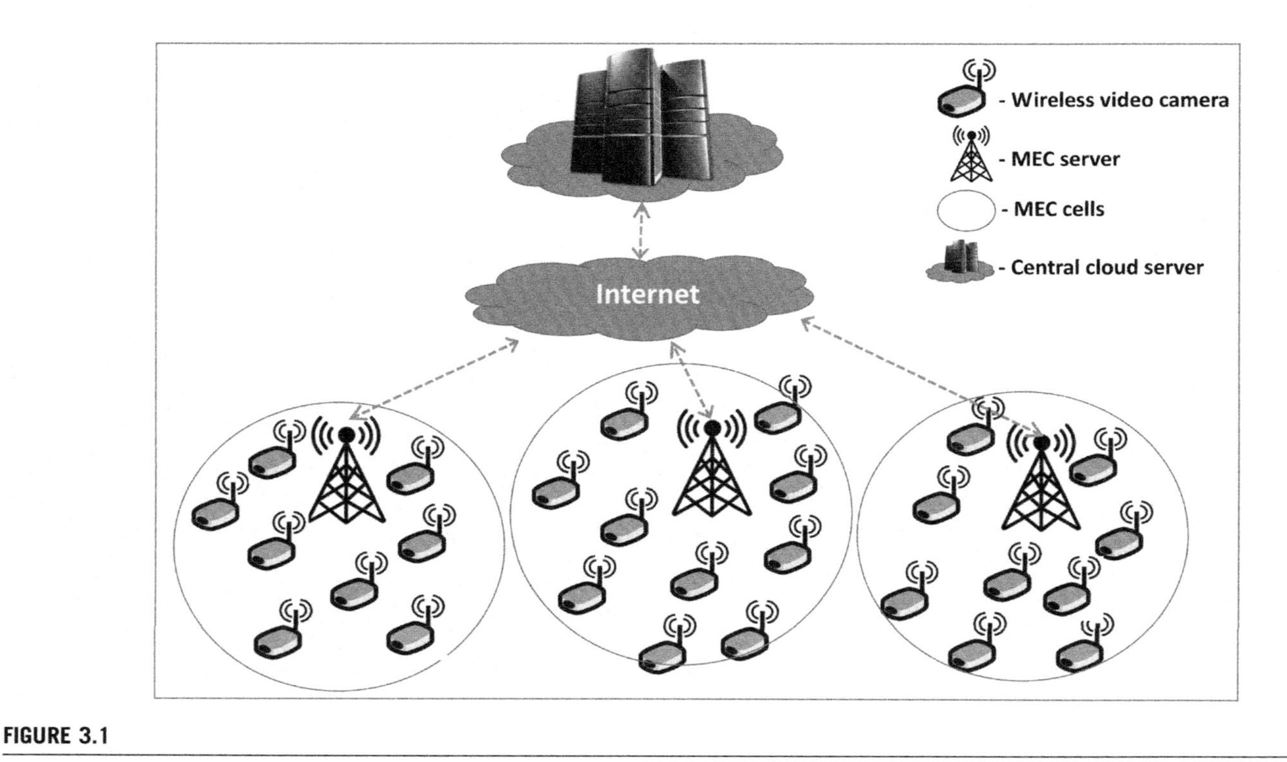


**FIGURE 3.1**

Overview of wireless surveillance framework.

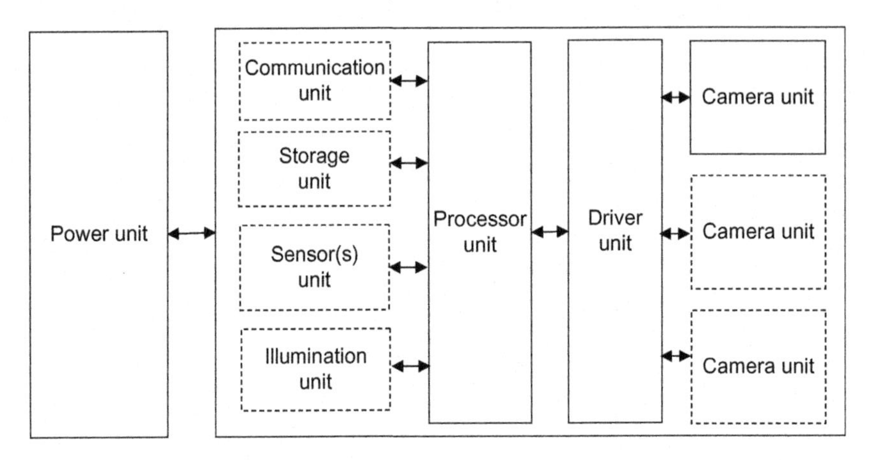


**FIGURE 3.2**

Smart wireless video camera—an overview.

has gained much importance, especially after the induction of the IPv6 addressing. A maximum of $2^{128}$ IoT devices could be connected to a network using the IPv6 addressing mode. The existing Internet infrastructures are low cost and ideal for ubiquitous connectivity [5]. Thus a WVC is considered as "a thing" in an IoT-based WSS. During the task assignment process in the IoT-based WSS, various factors including the capabilities and properties of the WVC, communication medium were not considered. To this end, a dynamic, self-aware task assignment for a wireless heterogeneous surveillance system is presented to reduce the energy consumption and maximize the usage of the wireless bandwidth.

The summary of the work is as follows. In Section 3.2, a study was carried out on: (1) the existing IoT-based WSS; and (2) the major factors influencing the task assignment. Section 3.3 presents the proposed dynamic, self-aware task assignment for a wireless heterogeneous surveillance system. The simulation analyzes and results of the proposed self-aware dynamic task assignment (SADTA) with the corresponding benchmark method are discussed in Section 3.4. Concluding remarks are provided in Section 3.5.

## 3.2 RELATED WORKS

The recent development in the IoT-based WSS is presented in this section. Colistra et al. [9] proposed a distributed consensus decision of homogeneous resource allocation among all the nodes in the network. Gossip and broadcast communication schemes were simulated to arrive at a consensus. The homogeneous resource utilization in a heterogeneous environment confines the performance of the network to a device with the least performance.

Khalil et al. proposed a game theory approach for two frameworks based on the evolutionary heuristic algorithm with a single objective to minimize energy consumption in the IoT applications.

They were: (1) a typical application framework was proposed to minimize energy consumption and maximize network lifetime; and (2) a crucial application framework was proposed to minimize energy consumption and maximize stability period.

Kokkonis et al. [1] proposed a network adaptive transmission protocol for a dynamic, multisensor-enabled surveillance system. Three separate dedicated channels were used for video, audio, and multisensor in the WSS. The frame rate transmission was adjusted dynamically to avoid network congestion. The protocol was found to be an ideal candidate for Wi-Fi and 4G networks. However, the heterogeneous properties were not considered.

Alsmirat et al. [5] proposed the IoT-based multitier framework for a wireless video surveillance system that aimed to minimize the high bandwidth demand and large storage requirement. The system used the MEC and cloud computing technologies. The MEC technology is ideal in data collection and bandwidth maintenance, while cloud computing offers unlimited storage and computing capabilities. The inherent nature and the properties were not considered while assigning tasks to the WVCs.

Bharti and Pattanaik [10] proposed a task requirement aware preprocessing and scheduling mechanism in the gateway to identify sensor nodes for the upcoming task. Tasks were classified based on spatial and temporal requirements such as negotiable and non-negotiable tasks. The properties of the things were not considered. The existing WSN task assignment methods, such as task allocation negotiation algorithm (TAN) [11] and self-organization based cooperative task allocation (SOCTA) [12], were modeled to cater the needs of a WSN. However, the existing WSN task assignment cannot be employed, as the data type, WVC configurations, and mission-critical nature of the WSS have to be considered.

The investigations revealed that the existing task assignment in the IoT and WSS did not consider the major properties of the edge nodes in the WSS during task assignment. The major factors influencing the task assignment, such as the properties of a WVC, wireless communication medium were not considered. A brief discussion on the factors is as follows.

### 3.2.1 FACTORS AFFECTING THE WIRELESS SURVEILLANCE SYSTEM

The major factors influencing the WSS are listed below.

*Video resolution* [4,5]: The number of pixels in the horizontal and vertical axes of a frame. The resolution of a video is represented by $H \times V$, where $H$ and $V$ are the number of horizontal and vertical pixels, respectively.

*Frame rate* [5]: Frame rate is the rate of change of frames on a display. It is measured as the number of frames displayed per second.

*Wireless bandwidth* [5]: Bandwidth is the maximum data that could be transmitted in a given time and is measured in bits per second. In a WSS, the global bandwidth is the maximum bandwidth between the intermediate server and the cloud servers. The local bandwidth is the maximum bandwidth between the WVC and its corresponding intermediate servers.

*Remaining energy* [13]: A battery-powered WVC is a highly energy-constrained device. The remaining energy of the WVC is expressed in Joules.

*Location* [13]: Based on the significance, a location could be broadly classified as a critical zone (CZ) and noncritical zone (NCZ) [13]. A CZ mandates persistent surveillance of the WVCs. In an NCZ, a set of WVCs is assigned to the surveillance task, while the rest of the WVCs are in a standby mode.

*Duration* [14]: WVCs could be dynamically assigned to be in a standby mode during the non-peak hours while fully operational during peak hours. A WMC's IR illumination may not be required in the bright daytime duration, in turn reducing the energy requirement in that particular span of time.

In summary, the literature survey revealed a lack of a novel dynamic task assignment algorithm for WSS. A dynamic task assignment algorithm for WSS, considering the property of the edges and factors affecting task assignment, is modeled in Section 3.3.

## 3.3 SELF-AWARE DYNAMIC TASK ASSIGNMENT ALGORITHM

Task assignment is the process of assigning a task to an ideal device [11,15]. The process of surveillance is decomposed into a set of indivisible tasks. The tasks are modeled as a directed acyclic graph as in [11], and the tasks are assumed to be independent of each other. The objectives of the task assignment are to: (1) maximize the lifetime of the WVCs; (2) improve the resolution of videos; and (3) minimize the bandwidth consumption. The aforementioned objectives of the task assignment in a WSS can be modeled as a multi-objective problem. The proposed SADTA algorithm is designed for WSS, and the WSS framework is as follows.

### 3.3.1 WIRELESS SURVEILLANCE SYSTEM FRAMEWORK

The overall framework of the proposed IoT-based WSS is depicted in Fig. 3.3. The primary participants in a WSS are: (1) the centralized cloud server (CCS); (2) MECSs; and (3) WVC. The CCS is assumed to have unlimited processing and storage capabilities. The $MECS_i$ $(1 \le i \le m;)$ are considered as powerful intermediate devices in the edge of the network performing local data aggregation and processing. The $WVC_{i,j}$ $(1 \le i \le m; 1 \le j \le n)$ are the leaf nodes of the WSS, connected wirelessly with its corresponding $MECS_i$.

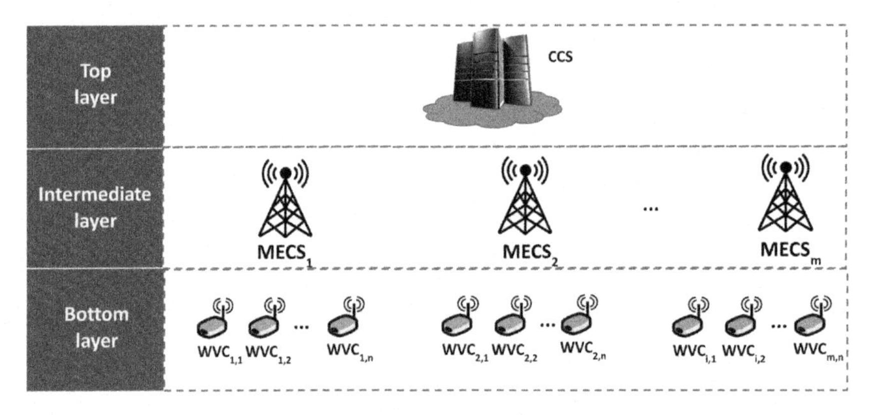


FIGURE 3.3

Self-aware dynamic task assignment framework.

The WVC's configuration and sensing units are processed by the MEC and CCS. The proposed task assignment occurs at the intermediate and top layers of the structure; however, the local cell task assignment is presented. The tasks are assigned to a WVC after the evaluation of the parameters using a decision support algorithm called the technique for order of preference by similarity to ideal solution (TOPSIS) method in the MEC.

### 3.3.2 TECHNIQUE FOR ORDER OF PREFERENCE BY SIMILARITY TO IDEAL SOLUTION

TOPSIS is a multicriteria decision analysis method based on the shortest geometric distance from the positive ideal solution and the longest geometric distance from the negative ideal solution [16]. A wide range of applications employs the TOSIS to solve the multicriteria problem [17]. The TOPSIS from [16] is adapted for the SADTA and is illustrated below.

*Step 1*: Generate the $n \times k$ evaluation matrix "$E$," where $e_{j,k}$ is the matrix element of the matrix "$E$." The matrix element $e_{j,k}$ is the performance metric of the $\mathrm{WVC}_{i,j}(1 \leq i \leq m)(1 \leq j \leq n)$ having $C_k(1 \leq k \leq p)$ criteria.

*Step 2*: Calculate the normalized decision matrix "$F$," where $f_{j,k}$ is the matrix element of the matrix "$N$." The value of $f_{ij}$ is calculated by Eq. (3.1), where $j = 1, 2, \ldots n;\ k = 1, 2, \ldots, p$.

$$f_{j,k} = e_{j,k} / \sqrt{\sum\nolimits_{k=1}^{p} \left(e_{j,k}\right)^2} \tag{3.1}$$

*Step 3*: Calculate the weighted normalized decision matrix "$G$," where $v_{j,k}$ is the matrix element of matrix "$G$." The matrix is obtained using Eq. (3.2), where $W$ is the criteria weighted matrix containing $w_k$ elements and $\sum_{k=1}^{p} w_k = 1, k = 1, 2, \ldots, p$.

$$G = W \times F \tag{3.2}$$

*Step 4*: Calculate the best $\mathrm{WVC}_b$ and the worst $\mathrm{WVC}_w$ using Eqs. (3.3) and (3.4), where $K_-$ and $K_+$ are the set of "$k$" criteria with negative and positive impacts.

$$\mathrm{WVC}_b = \left\{\left\langle \min\left(v_{j,k} | j = 1, 2, \ldots, n\right) | k \in K_- \right\rangle, \left\langle \max\left(v_{j,k} | i = 1, 2, \ldots, n\right) | k \in K_+ \right\rangle\right\} \tag{3.3}$$

$$\mathrm{WVC}_w = \left\{\left\langle \max\left(v_{j,k} | i = 1, 2, \ldots, n\right) | k \in K_- \right\rangle, \left\langle \min\left(v_{j,k} | i = 1, 2, \ldots, n\right) | k \in K_+ \right\rangle\right\} \tag{3.4}$$

*Step 5*: Calculate the L2 distance between the best WVC ($\mathrm{WVC}_b$) and the worst WVC ($\mathrm{WVC}_w$), where $d_{j,w}$ and $d_{j,b}$ are the distances from the luminaire "$l$" to the worst and best conditions, respectively.

$$d_{j,w} = \sqrt{\sum\nolimits_{j=1}^{n} \left(v_{ik} - v_{wj}\right)^2}; k = 1, 2, \ldots, p \tag{3.5}$$

$$d_{j,b} = \sqrt{\sum\nolimits_{j=1}^{n} \left(v_{ik} - v_{bj}\right)^2}; k = 1, 2, \ldots, p \tag{3.6}$$

*Step 6*: Calculate the relative closeness to the ideal solution.

$$\mathrm{CFV}_j = \frac{d_{j,w}}{d_{j,b} + d_{j,w}}; j = 1, 2, \ldots, n \tag{3.7}$$

At the end of the TOPSIS method, the camera fitness value (CFV)$(0 \leq \text{CFV} \leq 1)$ is calculated for every WVC. If the CFV is 1 or 0, the luminaire has the best or worst solution, respectively. The CFV perform the assigned illumination task corresponding to its role. The SADTA phases are discussed in the following.

### 3.3.3 SELF-AWARE DYNAMIC TASK ASSIGNMENT

The proposed SADTA is a centralized dynamic task assignment, and the phases are as follows: (1) registration; (2) self-evaluation; and (3) task evaluation phase.

1. *Registration phase*: A WVC upon joining a WSS registers to its nearest MECS with its corresponding properties, such as remaining energy, resolutions, frame rate, embedded sensors, and location, using a WVC_INFO$_{i,j}$ message.
2. *Self-evaluation phase*: The WVC gets triggered based on anyone of the following activities: (1) sensor actuation; (2) idleness; and (3) energy depletion. The WVC reports the activity to the concerned MEC, leading to the task evaluation phase by sending a TASKASSIGNMENT_INIT$_{i,j}$ message.
3. *Task evaluation phase*: The MECSs periodically perform assessments on their corresponding set of WVCs. The TOPSIS algorithm is used to calculate the overall CFV on a set of WVCs having "$p$" criteria, where $w_k$ is the corresponding weight of the $k$th criterion.

Thus the multiobjective problem of task assignment is converted into a single-objective problem by adapting the TOPSIS. The weights of the criteria are awarded based on the application needs. The tasks are assigned to the corresponding WVS with the highest CMV. The task assignment proposal is sent via a TASK_ASSIGNMENT_UPDATE message to the corresponding WVC. The corresponding WVC responds to the update message with a TASK_ASSIGNMENT_UPDATE_ACK$_{i,n}$ message.

The overview of the proposed phases is depicted in Fig. 3.4. The investigation of the proposed work with the corresponding benchmark method is presented in Section 3.4. The pseudocode for the algorithm is as follows.

**ALGORITHM 1. SADTA**

1: ***if*** $\boldsymbol{WVC_{i,j}}$***.status*** = *unregistered* ***then***
2: send *WVC − INFO*$_{i,j}$ to MECS$_i$
3: ***end if***
4: ***if*** $\boldsymbol{WVC_{i,j}}$***.TASKASSIGNMENT.Initiate*** = ***true***
5: send *TASK_ASSIGNMENT_INIT*$_{i,j}$ to MEC
6: ***end if***
7: Execute TOPSIS
8: Assign task to *WVC*$_{i,j}$ based on CFV; $j \in 1 \leq n$
9: Send *TASK_ASSIGNMENT_UPDATE*
10: ***for all*** $i \in WVC_{i,j}$ ***do***
11: send *ROLE_ASSIGNMENT_UPDATE_ACK* to MEC
12: ***end for***

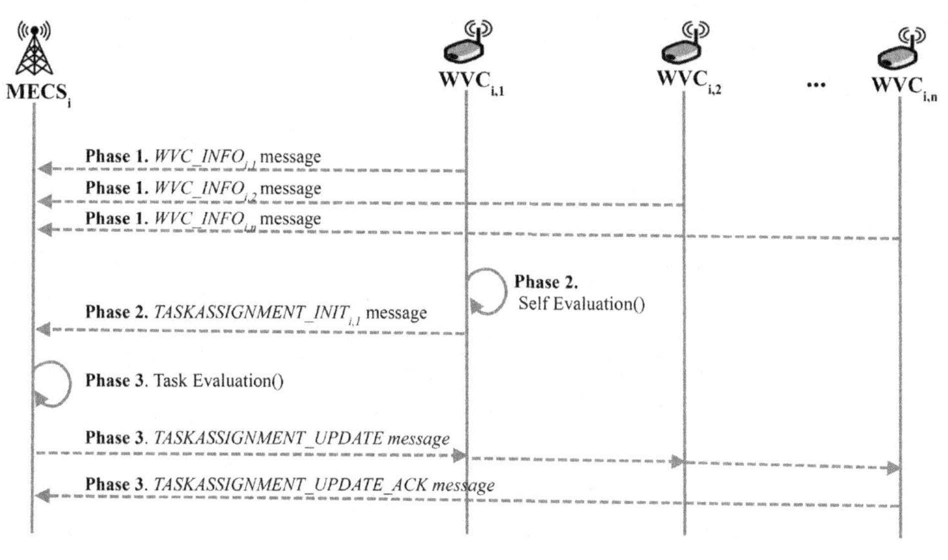


**FIGURE 3.4**

Overview of the task assignment phase.

## 3.4 SIMULATION ANALYSIS AND RESULTS

The simulations were carried out by using the SCI Lab software for the wireless surveillance scenario. The Waspmote, a wireless sensor node from Libelium, is taken as the WVCs in the wireless surveillance simulation. The Waspmote has one or more cameras and video boards. For simulation, a Waspmote with a single camera is taken. The heterogeneity in surveillance is achieved by initiating the WVC with random initial energy and equipping a few modes of resolution.

### 3.4.1 SIMULATION SETUP

The major parameters used in the simulation are tabulated in Table 3.1, and the rest is taken as in [5]. Every WVC is uniquely identified by its unique identifier and is connected to a MECS. A set of WVCs is connected to the nearest MECS, making a cell. The wireless connection within the cell is established by using Wi-Fi (IEEE 802.11g) connectivity.

The complete set of compatible video resolutions of the Waspmote is tabulated in Table 3.2. A total of six video resolutions could be generated by using the camera and video board.

The energy requirement of the individual component of a WVC is presented in Table 3.3. The following criteria were taken into consideration for the simulation: (1) bandwidth; (2) resolution; (3) frame rate; (4) remaining energy; (5) location; and (6) duration. A set of normalized weights is provided for the base SADTA method for the proposed method. Higher weight is awarded to the remaining energy criteria in the NCZ, while in a CZ scenario, higher weight is awarded to the location, resolution, and bandwidth criteria. The proposed method was compared with the local management with global weighted bandwidth distribution-enhanced (LM-WD) method presented in [5].

**Table 3.1 Simulation Parameters.**

| Properties | Values |
|---|---|
| MEC servers | 5 |
| Number of cells | 5 |
| Number of wireless cameras | Random (1–10) |
| Video frame rate | 15 fps |
| Wireless characteristics | IEEE 802.11g |
| Video compression | Raw |
| Wi-Fi Max bandwidth (local) | 54 Mbps |
| Wi-Fi Max bandwidth (global) | Random (4–100) Mbps |
| Simulation time | 24 h |

MEC, *Mobile edge computing.*

**Table 3.2 Waspmote Video Resolution.**

| Label | Resolutions (px) |
|---|---|
| R1 | $80 \times 48$ |
| R2 | $160 \times 120$ |
| R3 | $176 \times 144$ |
| R4 | $320 \times 240$ |
| R5 | $352 \times 288$ |
| R6 | $640 \times 480$ |

**Table 3.3 Energy Properties.**

| Properties | Value (J) |
|---|---|
| Initial energy | 23,760 |
| Camera | $158.5 \times 10^{-3}$ |
| IR LED | $75.48 \times 10^{-3}$ |
| IR sensor | $19.2 \times 10^{-3}$ |
| Presence sensor | $19.2 \times 10^{-3}$ |
| Wi-Fi send | $140 \times 10^{-3}$ |
| Wi-Fi receive | $140 \times 10^{-3}$ |

IR, *Infrared;* LED, *light emitting diode.*

### 3.4.2 BANDWIDTH ANALYSIS

The experimentation is enforcing various modes of video resolutions of WVCs for a fixed duration. The wireless bandwidth is continuously monitored by varying the WVC count and video resolution modes.

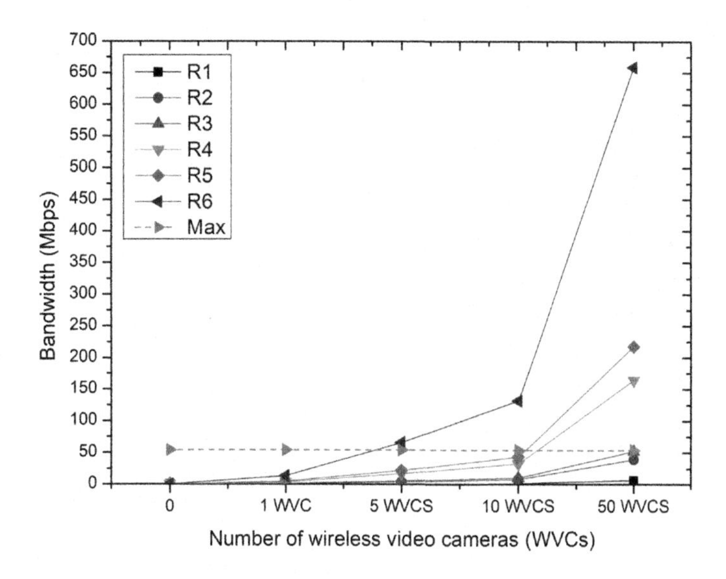


**FIGURE 3.5**

Bandwidth analysis.

Fig. 3.5 depicts the analysis performed on the total bandwidth used by the WVC based on the video resolutions. The dotted line represented by the label "Max" denotes the maximum permissible bandwidth of the IEEE 802.11(g). The investigation results revealed that the total bandwidth requirement of the WSS is directly proportional to the video resolution of the WVCs. The video packets transmitted above a cell's bandwidth would be dropped at the source. The simulation result reveals the need for an effective task assignment to effectively utilize the wireless bandwidth. The bandwidth analyzes of LM-WD and variants of SADTA were performed on the WVCs and summarized as hour-wise values.

Fig. 3.6 depicts the hour-wise total bandwidth usage of the proposed methods of SADTA and LM-WD-E. The bandwidth decreases considerably in the daytime, as the CZ, NCZ, and duration are taken into consideration in the proposed methods. In the SADTA scenario, both CZ and NCZ are considered, while in the SADTA-CZ and SADTA-NCZ scenarios, the entire location is considered as CZs and NCZs.

### 3.4.3 ENERGY CONSUMPTION

Fig. 3.7 depicts the total energy consumed by the WVCs in the WSS. The overall energy consumptions of the proposed SADTA-CZ and SADTA-NCZ with the LM-WD are 23.41% and 25.01%. The significant reduction of energy is due to the dynamic task assignment, considering the capabilities of the WVC, zones, and duration. The major part of the energy of a WVC is used by the camera and communication unit. In CZs, all the WVCs are on surveillance. The considerable drop is due to the dynamic illumination in the WVC. In NCZs, selected WVCs are dynamically assigned in a standby mode, awaiting for any event changes. Upon encountering any event, the onboard sensors trigger the camera unit in the WVC and update the MECS.

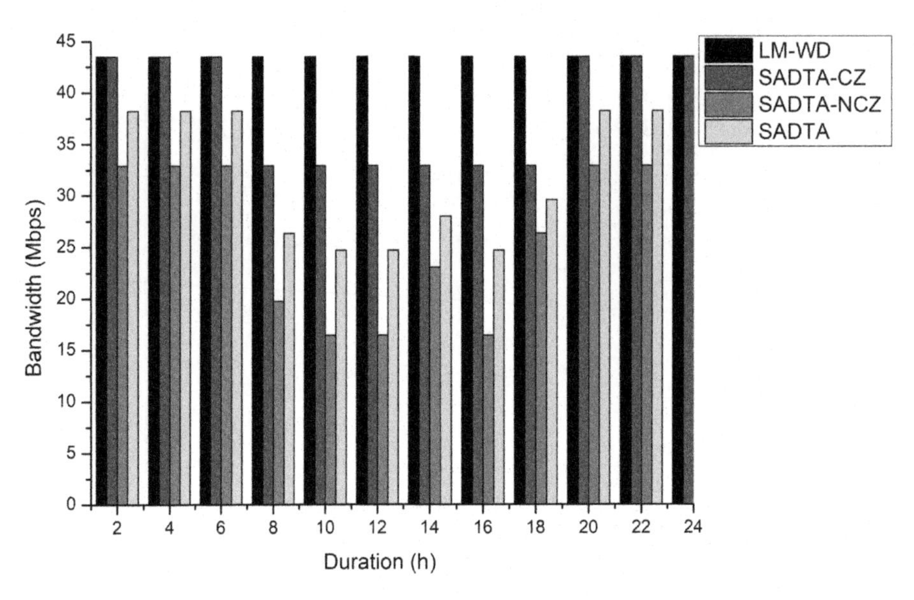


**FIGURE 3.6**

Hour-wise bandwidth utilization.

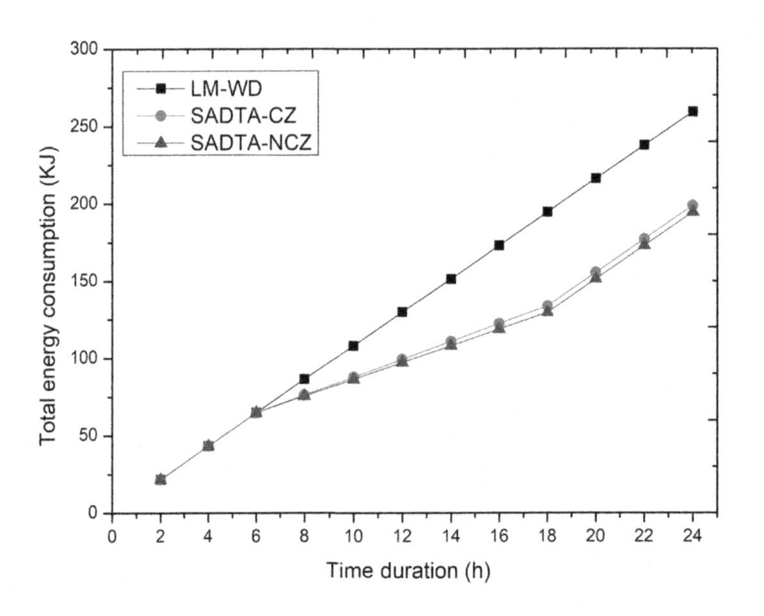


**FIGURE 3.7**

Total energy consumption in a day.

## 3.5 CONCLUSION

The IoT-based WSS are preferred over the legacy surveillance systems. Majorly, the existing task assignment algorithms in the literature were designed to optimize the bandwidth. The multiobjective need of the IoT-based WSS was addressed by the proposed SADTA algorithm. The various investigations, including dynamic resolution control, bandwidth utilization, and energy utilization, were performed on the WVCs by the corresponding MECSs with the corresponding benchmark method. The future work includes the investigation of the proposed task assignment on the global scenario with additional performance metrics.

## REFERENCES

[1] G. Kokkonis, K.E. Psannis, M. Roumeliotis, D. Schonfeld, Real-time wireless multisensory smart surveillance with 3D-HEVC streams for Internet-of-Things (IoT), J. Supercomput. 73 (3) (2017) 1044–1062.

[2] W.T. Chen, P.Y. Chen, W.S. Lee, C.F. Huang, (2008, May). Design and implementation of a real time video surveillance system with wireless sensor networks. In *VTC Spring 2008-IEEE Vehicular Technology Conference* (pp. 218–222). IEEE. (Marina Bay, Singapore 11–14 May 2008)

[3] M. Valera, S.A. Velastin, Intelligent distributed surveillance systems: a review, IEE P.-Vis. Image Sign. Process. 152 (2) (2005) 192–204.

[4] V.C. Banu, I.M. Costea, F.C. Nemtanu, I. Bădescu, Intelligent video surveillance system, in: IEEE 23rd International Symposium for Design and Technology in Electronic Packaging, Constanta, Romania, 2017, pp. 208–212.

[5] M.A. Alsmirat, Y. Jararweh, I. Obaidat, B.B. Gupta, Internet of surveillance: a cloud supported large-scale wireless surveillance system, J. Supercomput. 73 (3) (2017) 973–992.

[6] Waspmote Technical Guide, Libelium Comunicaciones Distribuidas S.L, 2018. [Online]. Available from: <http://www.libelium.com/downloads/documentation/waspmote_plug_and_sense_technical_guide.pdf> (Accessed 24.04.18).

[7] C. Liu, C. Fan, The design of remote surveillance system for digital family, in: Fifth International Conference on Intelligent Information Hiding and Multimedia Signal Processing, Kyoto, Japan, 2009, pp. 238–241.

[8] E.A. Khalil, S. Ozdemir, S. Tosun, Evolutionary task allocation in Internet of Things-based application domains, Futur. Gener. Comput. Syst. 86 (2018) 121–133.

[9] G. Colistra, V. Pilloni, L. Atzori, The problem of task allocation in the Internet of Things and the consensus-based approach, Comput. Netw. 73 (2014) 98–111.

[10] S. Bharti, K.K. Pattanaik, Ad hoc networks task requirement aware pre-processing and scheduling for IoT sensory environments, Ad Hoc Netw. 50 (2016) 102–114.

[11] V. Pilloni, P. Navaratnam, S. Vural, L. Atzori, R. Tafazolli, TAN: a distributed algorithm for dynamic task assignment in WSNs, IEEE Sens. J. 14 (4) (2014) 1266–1279.

[12] X. Yin, W. Dai, B. Li, L. Chang, C. Li, Cooperative task allocation in heterogeneous wireless sensor networks, Int. J. Distrib. Sens. Netw. 13 (10) (2017) 1–12.

[13] S. Misra, A. Vaish, Reputation-based role assignment for role-based access control in wireless sensor networks, Comput. Commun. (2011).

[14] T. Issac, S. Silas, E.B. Rajsingh, Luminaire aware centralized outdoor illumination role assignment scheme: a smart city perspective, in: J.D. Peter, A.H. Alavi, B. Javadi (Eds.), Advances in Big Data and Cloud Computing, Springer, Singapore, 2019, pp. 443–456.
[15] I. Titus, S. Silas, E.B. Rajsingh, Investigations on task and role assignment protocols in wireless sensor network, J. Theor. Appl. Inf. Technol. 89 (1) (2016) 209–219.
[16] M.S. García-cascales, M.T. Lamata, On rank reversal and TOPSIS method, Math. Comput. Model. 56 (5–6) (2012) 123–132.
[17] M. Johnson, S. Silas, Position aware and QoS based Service Discovery using TOPSIS for vehicular network, Int. J. Eng. Sci. Technol. 5 (03) (2013) 576–582.

CHAPTER

# SMART VEHICLE MONITORING AND TRACKING SYSTEM POWERED BY ACTIVE RADIO FREQUENCY IDENTIFICATION AND INTERNET OF THINGS

4

**D. Sugumar[1], T. Anita Jones[1], K.S. Senthilkumar[2], R.J.S. Jeba Kumar[1] and G. Thennarasi[1]**

*[1]Department of Electronics and Communication Engineering, Karunya Institute of Technology and Sciences, Coimbatore, India [2]Department of Computers and Technology, St. George's University, Grenada, West Indies*

## CHAPTER OUTLINE



## 4.1 RELATED WORKS

From the dawn of civilization, security has been evolving as a prime motive over our belongings. Organizations that are dependent on transport and travel relevance are at a high risk of security. In this modern world, it is essential that every travel company secures the right level of security to ensure that they are free from the threat of damage, danger, theft, or crime. Commercial and corporate risks require round the clock surveillance. Vehicle monitoring systems are employed to track the vehicle in order to get and process the monitoring data. With more cars on roads now than before, vehicle tracking has become an incredibly important tool for clients who wish to evade such a problematic havoc of insecurity. But, aside from avoiding traffic and road problems, the significant exercise is to monitor the vehicle with user-friendly environment that is welcomed by the present community.

Hoh et al. [1] proposed a typical traffic monitoring architecture with three entities: probe vehicles, GSM devices, and global positioning system (GPS) receivers. Chen and Liu [2] created the software integrated GPS, geographic information system (GIS), and GSM for intelligent vehicle monitoring system. Manihatty et al. [3], Lee et al. [4], Dhumal et al. [5], Kumar et al. [6], and

The Cognitive Approach in Cloud Computing and Internet of Things Technologies for Surveillance Tracking Systems.
DOI: https://doi.org/10.1016/B978-0-12-816385-6.00004-0

Security and Rengaraj [7] (child safety device) have also discussed the tracking facility with GPS-enabled devices. The major flip side of GPS-based tracking system is that, when the GPS device is switched off, the vehicle is out of monitoring/tracking control, and it can be driven unauthorized. Second, GPS signals can be blocked when the vehicle is in highly populated urban areas and underground tunnels. Kumar et al. furnished the monitoring system with GPS and GSM devices; however, it fails with the concept of secured remote monitoring capability. Hence, the major barrier of GPS is that it is prone to misuse by switching off the device. Moreover, GPS-based systems employed to monitor a vehicle's speed are normally costly to implement and to the subsequent maintenance costs.

Rahman et al. [8] established an e-plate radio frequency identification (RFID)-based tracking and management system. Hamzah et al. created an entry monitoring system with a passive RFID tag, only passing through the localized gate. It fails to monitor the vehicle continuously over its on-road journey. The use of passive RFID for tracking is limited to a short range of few centimeters, and it cannot be used for a real-time implementation [9]. Speed monitoring system [10], vehicle monitoring system [11], school bus tracking and security system [12], and transportation security system [13] were also developed based on RFID. Mainly, Hafeez et al. accomplished a speed monitoring system, but it fails with the distance of tracking zonal coverage. This system aims to calculate the speed of the vehicle with an active RFID, but there is no monitoring facility. Even, the car slot tracking facility with a passive RFID tag (vehicle monitoring system) discussed by Kiranmayi fails to cover a sufficient range. It has a range detection problem when the system ages. Dislocation of passive tags in the vehicle fails to track the RFID. Hence, the tracking is not practicable for security purpose.

In recent past, the Internet of Things (IoT) is a new technology that has attracted the attention of many researchers. The IoT applications include vehicular pollution monitoring [14], local-positioning awareness [15], bus monitoring system [16], intelligent vehicle monitoring system [17], dustbins [18], school bus tracking and arrival time prediction [19], smart school bus monitoring and notification system [20], children safety system [21], and intelligent bus positioning [22]. However, the IoT is implemented by using a combination of wireless sensor network (WSN)/RFID/GSM/GPRS technologies for these applications.

To avoid the problem of intrusion and hindrance, many alternative designs and techniques have been developed and implemented in the vehicles. Smart vehicle parking monitoring and management system [23] was developed by using high-end cameras and Android phones. However, this method cannot be used for monitoring and tracking, because its performance is limited by camera range. Many research efforts have been made to provide a standardized WSN/RFID integration framework to support smart vehicle monitoring and tracking systems. Zhang and Wang [24] have discussed about the opportunities and challenging problems of WSN/RFID integration. Sung et al. [25] and Ying and Kaixi [26] proposed to build a global standard infrastructure that combines the WSN and RFID.

In [27–29], the vehicle monitoring and identification systems are designed based on the IoT and RFID. Apart from these papers, in [30,31], a real-time car parking system and a device with identity verification for car drivers were presented, respectively. Prinsloo and Malekian [32] proposed a system to make use of the various solutions and opportunities provided by the IoT in order to give solutions to a real-world problem. They have used a combination of GPS, GSM, and RFID technologies for locating and tracking vehicles. They used the abilities of GPS technology as the communication platform in real time to collect data periodically, GSM technology for communication and storing data in the cloud, and RFID technology for sensing a module that is used in complex environments such as dense urban areas and underground tunnels. However, designing a vehicle security system and interfacing the monitoring, by the owners, viewable login webpage on a computer or a mobile phone will be the absolute solution to the current situation and the need. Notifications at anywhere can be observed by using a webpage that can be accessed by a computer as well as in a smartphone. Integrated engineering is a latest trend to solve problems. To be able to design a product using an integrated technology will be beneficial to any engineering problems and a huge contribution to the community. The summary of related works is shown in Table 4.1.

**Table 4.1 Summary of Related Works.**

| Technologies | Applications |
|---|---|
| GPS | Intelligent vehicle monitoring system [2] |
| | Vehicle tracking system [3] |
| | Vehicular tracking system [4] |
| | Vehicle tracking system [5] |
| | Vehicle monitoring and tracking system [6] |
| | Child safety device [7] |
| RFID | Plate number (e-plate): tracking and management system [8] |
| | Vehicle tracking system [9] |
| | Smart vehicle speed monitoring system [10] |
| | Vehicle monitoring system [11] |
| | School bus tracking and security system [12] |
| | Transportation security [13] |
| IoT | Vehicular pollution monitoring [14] |
| | Local-positioning awareness [15] |
| | Bus monitoring system [16] |
| | Intelligent vehicle monitoring system [17] |
| | Dustbins [18] |
| | School bus tracking and arrival time prediction [19] |
| | Smart school bus monitoring and notification system [20] |
| | Children safety system [21] |
| | Intelligent bus positioning [22] |
| Cameras + Android OS | Smart vehicle parking monitoring and management system [23] |
| RFID + WSN | Logistics management system [26] |
| RFID + IoT | Automatic vehicle identification system [27] |
| | Goods dynamic monitoring and controlling system [29] |
| | Real-time car parking system [30] and vehicle location system [32] |
| | Device with identity verification for car driving [31] |

GPS, *Global positioning system;* IoT, *Internet of Things;* WSN, *wireless sensor network.*

This chapter presents the design and implementation of vehicle tracking, vehicle monitoring, and vehicle status system (vehicle number and location with date and time at the respective zones). The designed system consists of a sensor actuator module and a communication module to acquire the input signals and monitor the vehicle using the active RFID and IoT technology. A dedicated, portable, cost-effective, and flexible vehicle tracking and monitoring system deployed into automobile. To demonstrate the feasibility and effectiveness of the proposed system, vehicle number, reserved vehicle zone, and duration of entry and exit with date and time are monitored by using the local zone map. Maps of grid sequence are vested on the zones. The indication in the zones also gives the details about the vehicle. The status notifications for vehicle entry to the zones, location, and duration are implemented with an integrated communication technology.

## 4.2 NEED FOR SMART VEHICLE MONITORING SYSTEM

Every year, thousands of vehicles are stolen on the roads and in the unsecured parking areas. Banks employ exclusive assistance for replenishing cash in automated teller machines (ATMs); however, intruders still attempt to steal the cash by opening the doors or breaking the glasses of the vehicle. Because of this, organizations are afraid to ship cash in the vehicles. In the course of rendering vehicles for the local travel purposes, the proprietor wants to know the current location of the vehicle. GPS-enabled devices are prevalent now for tracking and monitoring purposes. The main drawback of the GPS system is that, when a GPS-enabled device is switched off, the device is out of the tracking/coverage area; hence, it can be driven unauthorized. To avoid the havoc of intrusion and hindrance, there is a need for a smart vehicle monitoring system. This novel designed system with an active RFID and IoT technology is for a complete monitoring purpose. This novel smart vehicle monitoring system powered by an active RFID and IoT (SVM-ARFIoT) system that provides a viewable login webpage furnished with the monitoring facility will be the absolute solution for the needed secured monitoring system.

## 4.3 DESIGN OF SMART VEHICLE MONITORING SYSTEM

SVM-ARFIoT is utilized here, which is cost-effective and secured. The system gives tracking assistance over the connected devices. Hence, this system overcomes the issue of GPS, as it is mainly focused on the continuous monitoring with an active RFID and IoT technology. Hence, there is a need for a system that is more secured and cost-effective. The status notifications for vehicle entry to the zones, location, and duration are implemented with an integrated communication technology. The detailed block diagram is depicted in Fig. 4.1. ESP 8266 is used to push the data to the cloud, that is, the IoT domain. The harvested information from the on-field territory is a unique RFID number (binary), zone number (binary), and WSN number (binary) in the format of "1010;1;1". This pattern is pushed to the IoT platform. From the already stored vehicle information by the customer during lending, the vehicle number is noted, and it is mapped for the harvested RFID number. Thus the system can identify the vehicle in the particular zone, that is, monitoring is achieved with a low bandwidth.

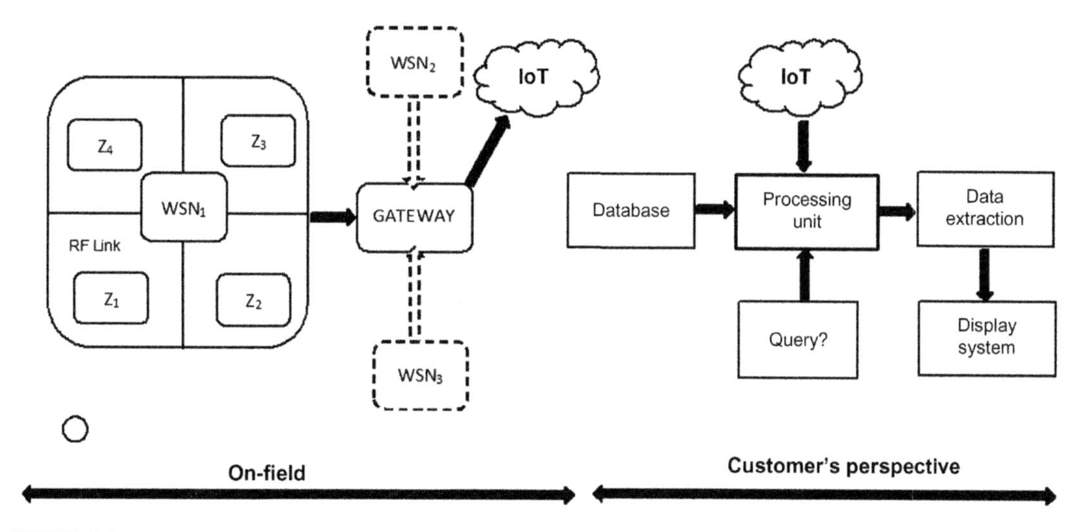


**FIGURE 4.1**

Detailed block diagram of the smart vehicle monitoring system powered by active radio frequency identification tag and Internet of Things.

The layout of the specified area is initially gridded into zones based on the range of coverage. The active RFID transmitters are equipped in a mobile fashion, and they have been housed in the vehicle. The RF wireless sensors serve as stationary active RFID receivers that are placed as per the range of detection. This geographical area is split into zones depending on the active RFID detection range. The stationary active RFID receivers are retained in the range specified zones. All these stationary active RFIDs are connected to form a WSN. Fig. 4.2 shows the zonal grid diagram of the SVM-ARFIoT, the yellow dot in the diagram indicates the vehicle entering the zone of WSN housed with an active RFID transmitter, and the blue dots indicate the sensor nodes attached with the receiver module of the WSN.

The functional flowchart of the SVM-ARFIoT is shown in Fig. 4.3. The functional flow of the system is divided into two phases: on-field (hardware phase) and customer perspective (software phase). The hardware phase computes the unique active RFID number and zone appending from each zone. The software phase matches the unique active RFID number with the vehicle number of the customer, which is stored in the database. A customer query is sent to the processing unit to monitor the specified vehicle. Database management system is used to store the customer vehicle data to correlate the active RFID number and vehicle number for monitoring purpose. Time stamp is given during each entry to the zone to have control over efficient monitoring. Table 4.2 compares the existing GPS-enabled automobile tracking system with the SVM-ARFIoT in various dimensions.

## 4.4 EVALUATION OF SVM-ARFIOT

The active RFID module works under an industrial, scientific and medical (ISM) frequency band of 433 MHz. It is furnished to act as a 4-bit data encoder (HT12E). This active RFID module emits a

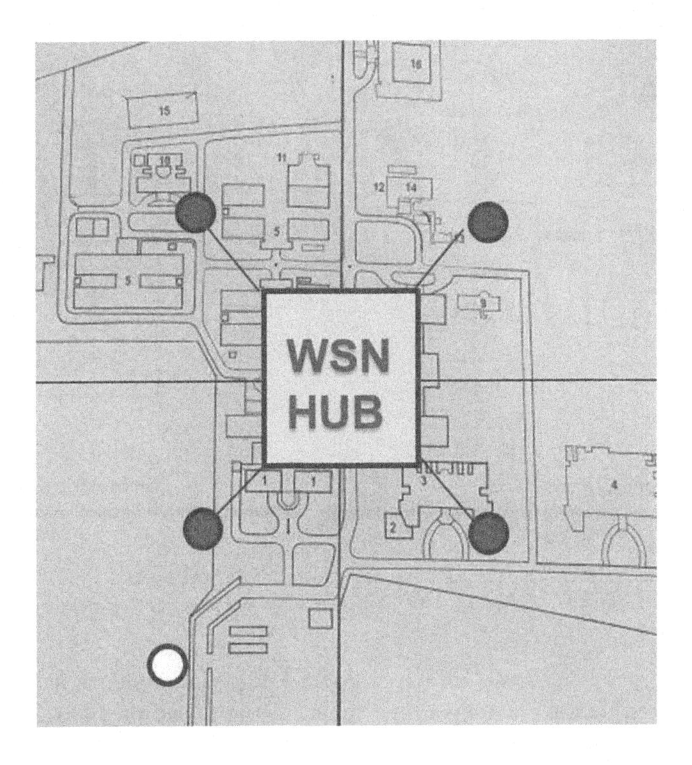


**FIGURE 4.2**

Zonal grid diagram of smart vehicle monitoring system powered by active radio frequency identification tag and Internet of Things.

4-bit data combinational code as 0000 to 1111, which we can have in 16 combinations and can be placed in 16 vehicles. These four bits can be scaled sufficiently to many by adding an additional encoding bit in the HT12E module. This flexibility of scaling in a low-cost fashion is a boon to many micro, small, and medium lending enterprises for their automobile monitoring. The active RFID module works under the operating voltage of 2.4–12 V. The 3 bit combinational code has to be made uniformly in both the transmitter and receiver sections to establish the proper encoding and decoding processes. Fig. 4.4 portrays the active RFID transmitter module that works in the ISM band (433 MHz) implanted on the automobile.

In the receiver part, we have the 4-bit decoder IC (HT12D) as an active RFID receiver. This active RFID receiver will receive the 4 bit, which is transmitted by the transmitter. It is received by the RF waves with 433 MHz. Fig. 4.5 portrays the active RFID receiver mounted on the WSN hub (MSP430) with the IoT transceiver module (ESP8266) to push the harvested data to the IoT domain.

Raspberry Pi 3 that has inbuilt-enabled Wi-Fi access, to act as a central server. All the harvested data are stored in this server, which is connected to all the nearby WSNs. Fig. 4.6 portrays the server setup of Raspberry Pi 3.

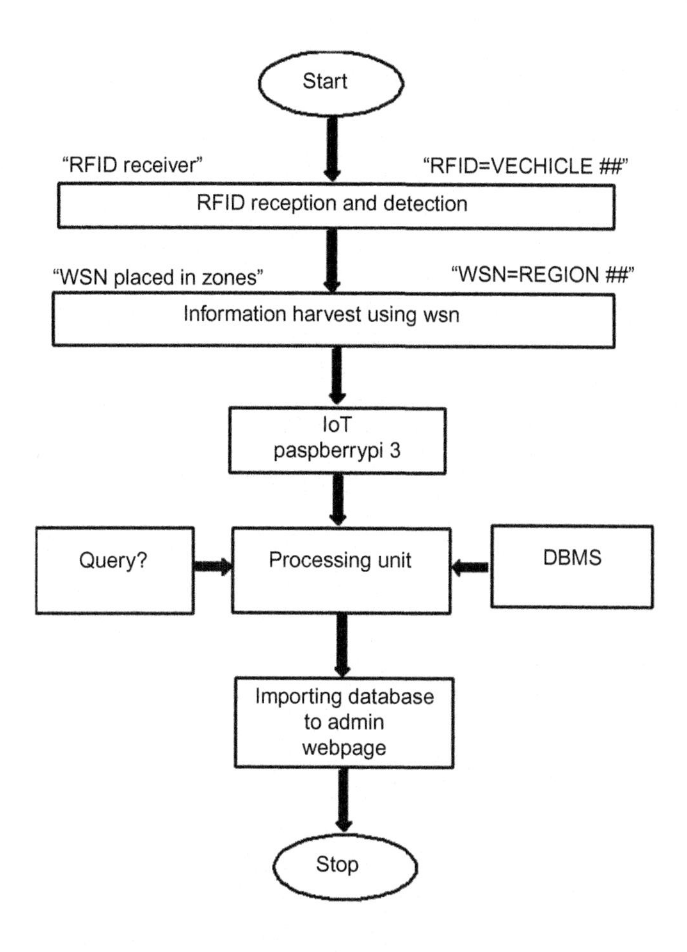


**FIGURE 4.3**

Flowchart of the smart vehicle monitoring system powered by active radio frequency identification tag and Internet of Things.

The server collects the data in the form of individual data bits and zone numbers, as portrayed in Fig. 4.7. The snapshot of the harvested data shown in the server is a putty screen connected via a serial port and secure shell. From Fig. 4.7, we can infer the data bits (d1-d4), and the zone numbers are harvested separately; hence, there is low possibility of error.

The real-time data received from the IoT server to the database of phpMyAdmin is portrayed in Fig. 4.8. A random parity bit is appended before each data bit (d1-d4) for security. Facilities for time stamp with date are included to have an efficient control for a secured monitoring system.

The IoT support login page is created by using the HTML code. A unique account is created for each customer by providing unique user name and password, as depicted Fig. 4.9. This feature of entry control mechanism is to provide security from an unauthorized person viewing the monitoring details.

**Table 4.2 Comparison of Conventional GPS-Based Vehicle Monitoring System Versus SVM-ARFIoT-Based Vehicle Monitoring System.**

| Conventional GPS-Based Vehicle Monitoring System | SVM-ARFIoT-Based Vehicle Monitoring System |
|---|---|
| GPS module and chipset are not cost-effective. | SVM-ARFIoT system is cost-effective. |
| It demands a huge bandwidth, as it is solely depending on software computation. | It needs a low bandwidth. Balance of computational load is equally divided into both hardware and software sections. |
| It is not scalable. | It is highly scalable, as the codes can be reused. |
| It is prone to hacking. | It is immune to hacking. Parity codes are appended before the unique active RFID number for security. |
| It is directly done on the basis of GPS module integration to a network. | It is accomplished with smart integration of active RFID, and the WSN is connected to the IoT technology to enable the best data extraction and efficient handling even. |

GPS, *Global positioning system;* SVM-ARFIoT, *smart vehicle monitoring system powered by active radio frequency identification tag and Internet of Things;* WSN, *wireless sensor network.*

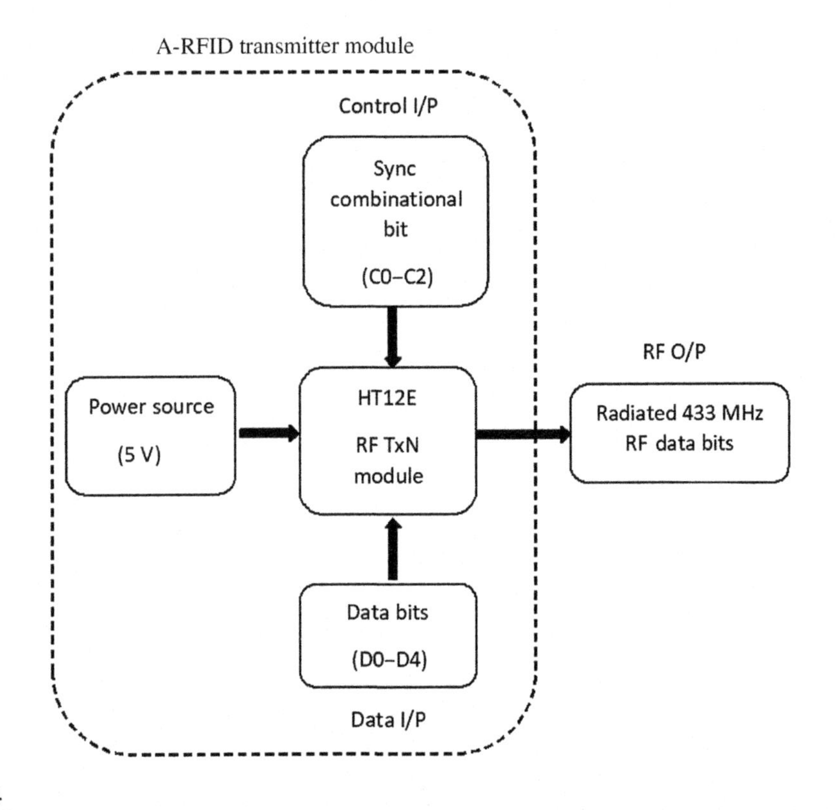


**FIGURE 4.4**

Active radio frequency identification transmitter module implanted on the automobile.

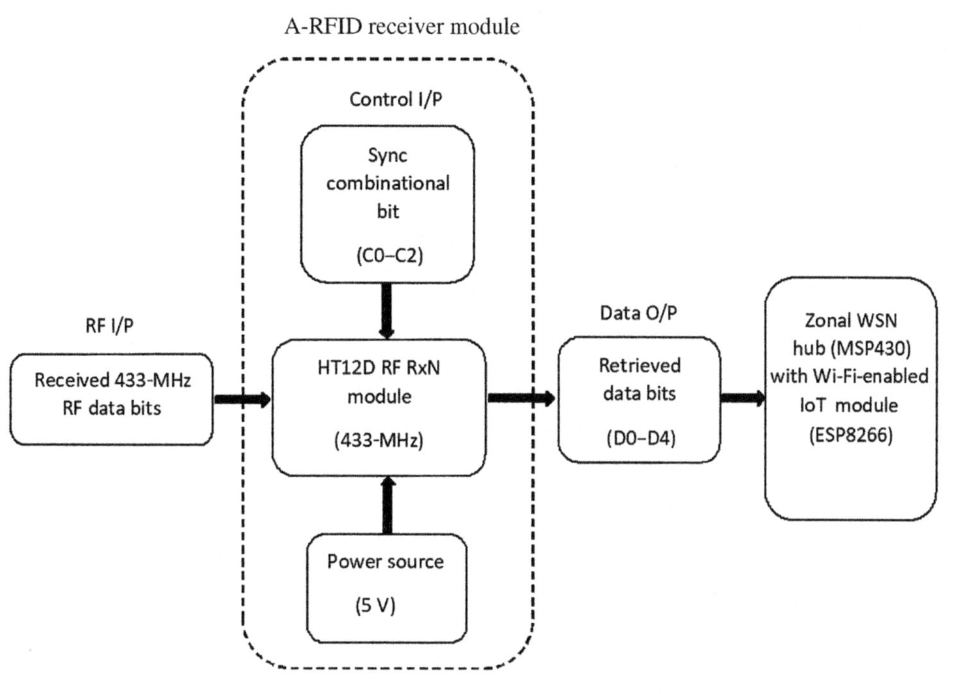


**FIGURE 4.5**

Active RFID receiver of 433 MHz mounted on the WSN hub (MSP430) with the IoT transceiver module (ESP8266).

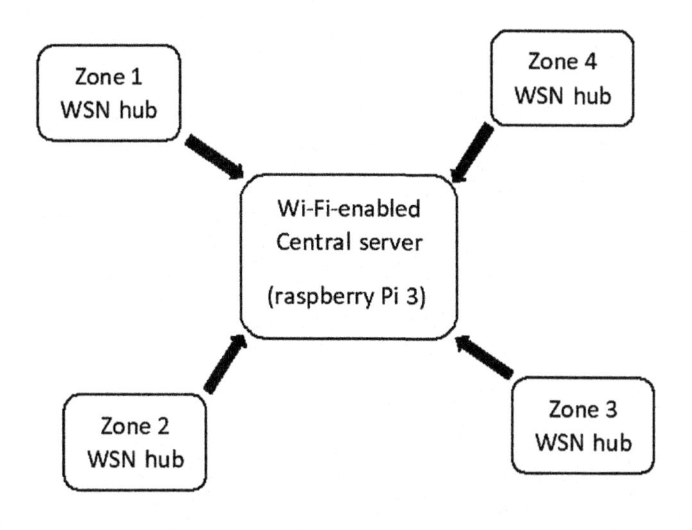


**FIGURE 4.6**

Raspberry Pi 3 with Wi-Fi-enabled access.

**Captured Data at T1 Seconds:**
"d1": 0 "d2": 0 "d3": 0 "d4": 1 "Zone": "Zone 1"

**Captured Data at (T1 + 30) Seconds:**
"d1": 0 "d2": 0 "d3": 0 "d4": 1 "Zone": "Zone 1"

**Captured Data at (T1 + 60) Seconds:**
"d1": 0 "d2": 0 "d3": 1 "d4": 0 "Zone": "Zone 3"

**Captured Data at (T1 + 90) Seconds:**
"d1": 0 "d2": 0 "d3": 1 "d4": 0 "Zone": "Zone 3"

**T1: Time stamp; d1–d4: Unique active RFID bits; Zone: Topographical vehicle position.**

**FIGURE 4.7**

Raspberry Pi 3 harvested data from the wireless sensor network shown in the putty window.

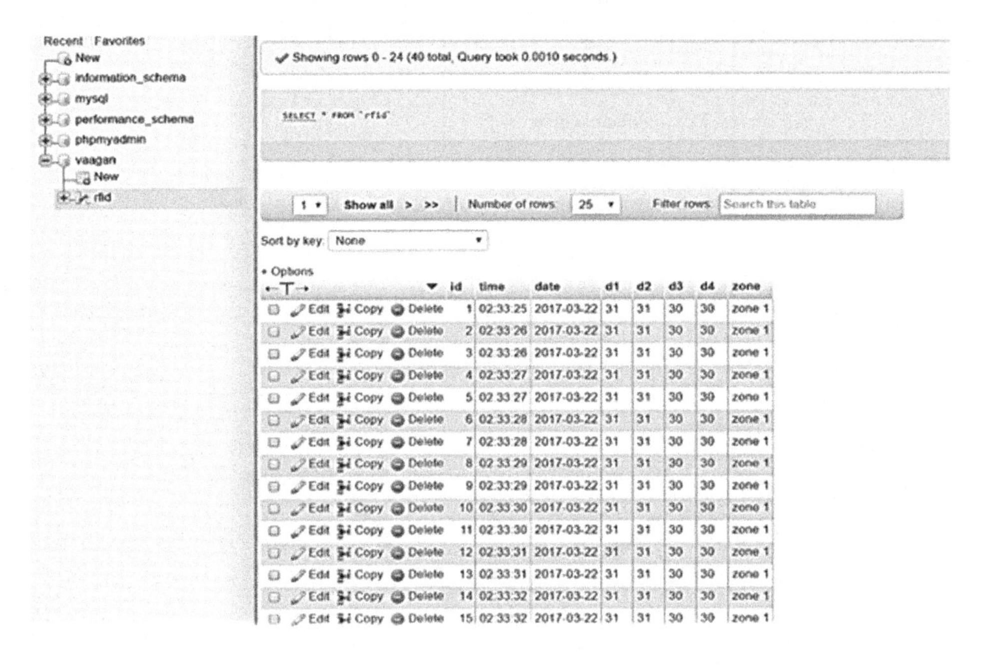


| | id | time | date | d1 | d2 | d3 | d4 | zone |
|---|---|---|---|---|---|---|---|---|
| Edit Copy Delete | 1 | 02:33:25 | 2017-03-22 | 31 | 31 | 30 | 30 | zone 1 |
| Edit Copy Delete | 2 | 02:33:26 | 2017-03-22 | 31 | 31 | 30 | 30 | zone 1 |
| Edit Copy Delete | 3 | 02:33:26 | 2017-03-22 | 31 | 31 | 30 | 30 | zone 1 |
| Edit Copy Delete | 4 | 02:33:27 | 2017-03-22 | 31 | 31 | 30 | 30 | zone 1 |
| Edit Copy Delete | 5 | 02:33:27 | 2017-03-22 | 31 | 31 | 30 | 30 | zone 1 |
| Edit Copy Delete | 6 | 02:33:28 | 2017-03-22 | 31 | 31 | 30 | 30 | zone 1 |
| Edit Copy Delete | 7 | 02:33:28 | 2017-03-22 | 31 | 31 | 30 | 30 | zone 1 |
| Edit Copy Delete | 8 | 02:33:29 | 2017-03-22 | 31 | 31 | 30 | 30 | zone 1 |
| Edit Copy Delete | 9 | 02:33:29 | 2017-03-22 | 31 | 31 | 30 | 30 | zone 1 |
| Edit Copy Delete | 10 | 02:33:30 | 2017-03-22 | 31 | 31 | 30 | 30 | zone 1 |
| Edit Copy Delete | 11 | 02:33:30 | 2017-03-22 | 31 | 31 | 30 | 30 | zone 1 |
| Edit Copy Delete | 12 | 02:33:31 | 2017-03-22 | 31 | 31 | 30 | 30 | zone 1 |
| Edit Copy Delete | 13 | 02:33:31 | 2017-03-22 | 31 | 31 | 30 | 30 | zone 1 |
| Edit Copy Delete | 14 | 02:33:32 | 2017-03-22 | 31 | 31 | 30 | 30 | zone 1 |
| Edit Copy Delete | 15 | 02:33:32 | 2017-03-22 | 31 | 31 | 30 | 30 | zone 1 |

**FIGURE 4.8**

Real-time data received from the Internet of Things domain to the database with date, time-stamp, and encoded data bits.

The monitoring admin page consists of various functionalities like import database from the IoT server, list of registered vehicles, specific option to delete the particular vehicle's outdated details from the main database permanently, the capability to monitor the vehicle by providing an active RFID number in the monitor tab, and the ability to refresh the entire database. All these functionalities are portrayed in the admin webpage of Fig. 4.10.

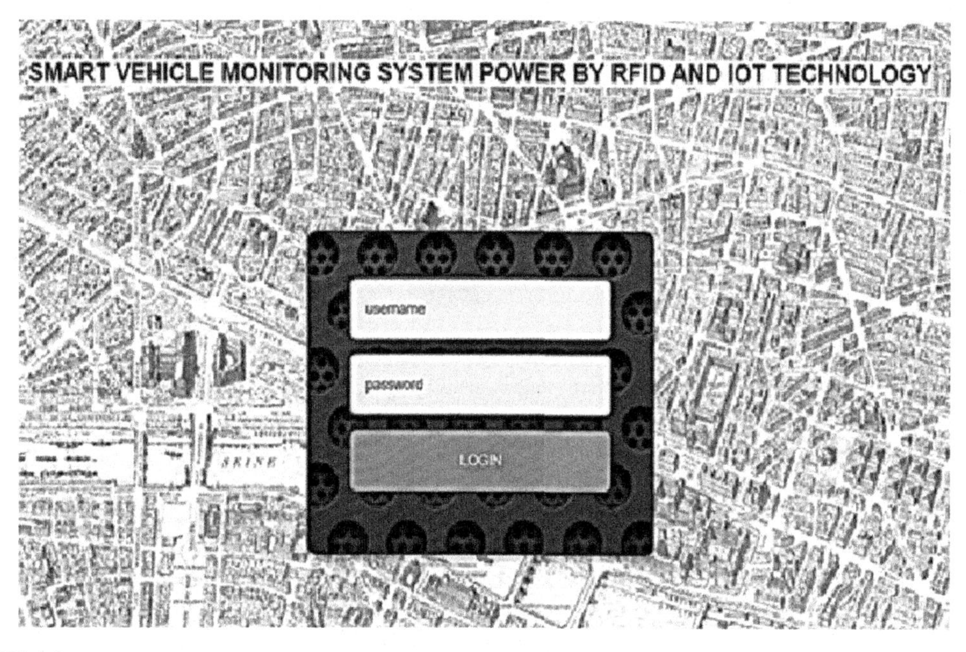


**FIGURE 4.9**

Login page of SVM-ARFIoT.

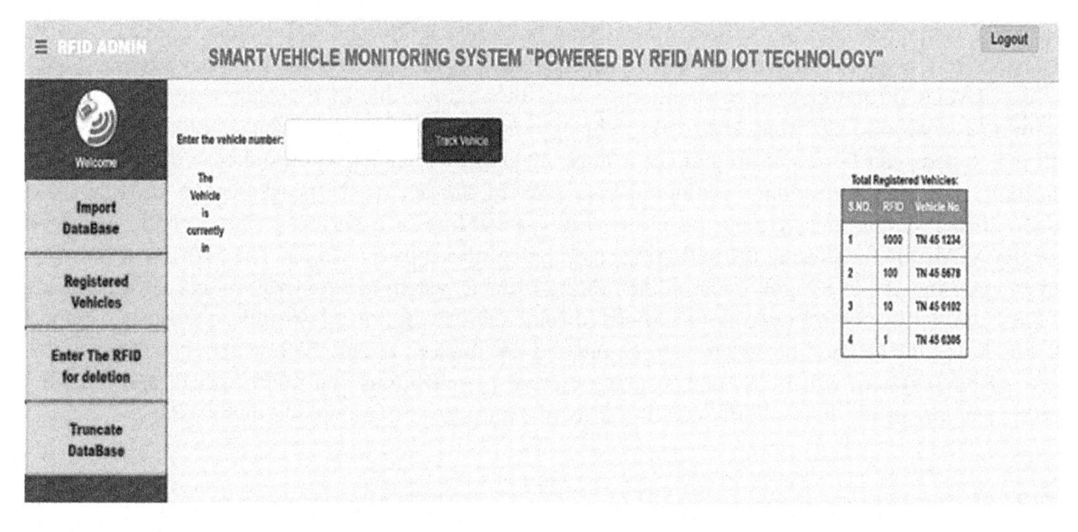


**FIGURE 4.10**

Admin monitoring page with multiple options for smart vehicle monitoring system powered by active radio frequency identification tag and Internet of Things.

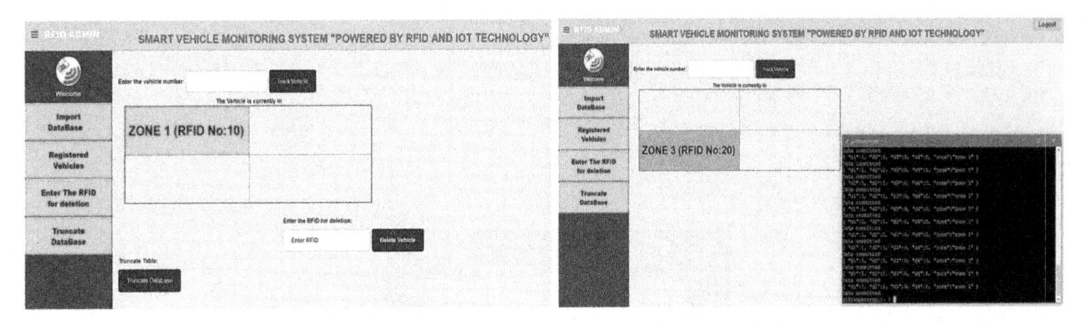


**FIGURE 4.11**

Tracking page with multiple options for smart vehicle monitoring system powered by active radio frequency identification tag and Internet of Things.

Upon the successful entry of RFID in the "Track Vehicle" search box, the zone of the present location of that particular active RFID-holding vehicle is shown in the zone map. Fig. 4.11 shows the monitored results of the particular vehicle with the active RFID number of 0010, and the results are successfully reflected in the SVM-ARFIoT webpage support for customers. For the purpose of research, the zones have been indicated as matrices; however, in real time, a GIS image can be used as an alternative to indicate the accurate location of the vehicle.

## 4.5 CONCLUSION

The smart vehicle monitoring system powered by an active RFID and IoT technology proposed in this chapter is a cost-effective and highly secured system compared with the conventional GPS-enabled devices for the purpose of monitoring. Real-time monitoring of a vehicle is displayed in the appropriate zones in the front end, IoT webpage support for customer successfully. This SVM-ARFIoT system can be scaled to support a huge number of vehicles by encoding techniques, code reusability concept, and adding additional data pins in the active RFID transceiver module. The SVM-ARFIoT monitoring system will ensure safe onboarding of a currency chest vehicle to safely load the ATMs, by monitoring the authorized road path utilized by the driver. The SVM-ARFIoT system can be extended to a highly secured monitoring official system to avoid victim driving misshapen by terrorist, that is, control over an unauthorized route for driving for a law-guilty person in a police vehicle during the official journey toward an ordered destination, to enforce law correctly. Hence, the SVM-ARFIoT system will be a boon to many official organizations and micro, small, and medium lending enterprises for their secured and cost-effective purposes of automobile monitoring.

## REFERENCES

[1] B. Hoh, M. Gruteser, H. Xiong, A. Alrabady, Enhancing security and privacy in traffic-monitoring systems, IEEE Pervasive Comput. 5 (4) (2006) 38–46.

[2] P. Chen, S. Liu, Intelligent vehicle monitoring system based on GPS, GSM and GIS, in: WASE International Conference on Information Engineering, IEEE Publisher, 2010, pp. 38–40.
[3] S. Lee, G. Tewolde, J. Kwon, Design and implementation of vehicle tracking system using GPS/GSM/GPRS technology and smartphone application, in: IEEE World Forum on Internet of Things, IEEE Publisher, 2014, pp. 353–358.
[4] T. Manihatty, B. Umamaheswaran, R. Kumar, V.M. Bojan, Designing vehicle tracking system—an open source approach, in: IEEE International Conference on Vehicular Electronics and Safety, IEEE Publisher, 2014, pp. 135–140.
[5] A. Dhumal, A. Naikoji, Y. Patwa, M. Shilimkar, M.K. Nighot, Vehicle tracking system using GPS and Android OS, Int. J. Adv. Res. Comput. Eng. Technol. 4 (4) (2015) 1220–1224.
[6] B.H. Kumar, S.F. Tehseen, S. Thanveer, G.V. Krishna, S.M. Akram, Vehicle monitoring and tracking system using GPS and GSM technologies, Int. Res. J. Eng. Technol. 3 (4) (2016) 72–74.
[7] C. Security, V. Rengaraj, A study and implementation of Smart ID card with M-learning and child security, in: 2nd International Conference on Applied and Theoretical Computing and Communication Technology, IEEE Publisher, 2016, pp. 305–311.
[8] T.A. Rahman, S. Kamal, A. Rahim, RFID vehicle plate number (e-Plate) for tracking and management system, in: 2013 International Conference on Parallel and Distributed Systems, IEEE Publisher, 2013, pp. 611–616.
[9] M.P. Hamzah, N.M.M. Noor, M.N. Hassan, N.F.A. Mamat, M.A.S.M. Rifin, Implementation of vehicle tracking using radio frequency identification (RFID): vTrack, Int. J. Digit. Content Technol. Appl. 7 (16) (2013) 38–45.
[10] F. Hafeez, M. Al Shammrani, O. Al Shammary, Smart vehicles speed monitoring system using RFID, Int. J. Adv. Res. Electr. Electron. Instrum. Eng. 4 (4) (2015) 1860–1864.
[11] D. Kiranmayi, Vehicle monitoring system using RFID, Int. J. Comput. Sci. Inf. Technol. 7 (3) (2016) 1444–1447.
[12] S. Shah, B. Singh, RFID based school bus tracking and security system, in: International Conference on Communication and Signal Processing, IEEE Publisher, 2016, pp. 1481–1485.
[13] S.R. Singh, Sensors in transportation for security using RFID, in: Second International Conference on Electronics, Communication and Aerospace Technology, IEEE Publisher, 2018, pp. 714–720.
[14] S. Manna, S. Bhunia Sanka, M. Nandini, Vehicular pollution monitoring using loT, in: IEEE International Conference on Recent Advances and Innovations in Engineering, IEEE Publisher, 2014, pp. 1–5.
[15] F. Kirsch, R. Miesen, M. Vossiek, Precise local-positioning for autonomous situation awareness in the Internet of Things, in: IEEE MTT-S International Microwave Symposium, IEEE Publisher, 2014, pp. 1–4.
[16] J. Zambada, R. Quintero, R. Isijara, R. Galeana, L. Santillan, An IoT based scholar bus monitoring system, in: IEEE First International Smart Cities Conference, IEEE Publisher, 2015, pp. 1–6.
[17] R.B. Pendor, P.P. Tasgaonkar, An IoT framework for intelligent vehicle monitoring system, in: International Conference on Communication and Signal Processing, IEEE Publisher, 2016, pp. 1694–1696.
[18] S. Mirchandani, R. Joseph, IoT enabled dustbins, in: International Conference on Big Data, IoT and Data Science, IEEE Publisher, 2017, pp. 73–76.
[19] R.C. Jisha, A. Jyothindranath, L.S. Kumary, IoT based school bus tracking and arrival time prediction, in: International Conference on Advances in Computing, Communications and Informatics, IEEE Publisher, 2017, pp. 509–514.
[20] J.T. Raj, J. Sankar, IoT based smart school bus monitoring and notification system, in: IEEE Region 10 Humanitarian Technology Conference, IEEE Publisher, 2017, pp. 89–92.

[21] L.D. Errico, F. Franchi, F. Graziosi, C. Rinaldi, F. Tarquini, Design and implementation of a children safety system based on IoT technologies, in: 2nd International Multidisciplinary Conference on Computer and Energy Science, IEEE Publisher, 2017, pp. 1−6.
[22] X. Feng, J. Zhang, J. Chen, G. Wang, L. Zhang, R. Li, Design of intelligent bus positioning based on Internet of Things for smart campus, IEEE Access 6 (2018) 60005−60015. 2018.
[23] M.Y. Aalsalem, W.Z. Khan, CampusSense—a smart vehicle parking monitoring and management system using ANPR cameras and android phones, ICACT Trans. Adv. Commun. Technol. 5 (2) (2016) 809−815.
[24] L. Zhang, Z. Wang, Integration of RFID into wireless sensor networks: architectures, opportunities and challenging problems, in: Proceedings of the Fifth International Conference on Grid and Cooperative Computing Workshops, IEEE Publisher, 2006, pp. 463−469.
[25] J. Sung, T.S. Lopez, D. Kim, The EPC sensor network for RFID and WSN integration infrastructure, in: Fifth Annual IEEE International Conference on Pervasive Computing and Communications Workshops, IEEE Publisher, 2007, pp. 618−621.
[26] W. Ying, W. Kaixi, The Building of logistics management system using RFID and WSN technology, in: IEEE International Conference on Computer Science and Automation Engineering, IEEE Publisher, 2008, pp. 651−654.
[27] M. Yu, D. Zhang, Y. Cheng, M. Wang, An RFID electronic tag based automatic vehicle identification system for traffic IOT applications, in: Chinese Control and Decision Conference, IEEE Publisher, 2011, pp. 4192−4197.
[28] X. Jia, Q. Feng, T. Fan, Q. Lei, RFID technology and its applications in Internet of Things (IoT), in: 2nd International Conference on Consumer Electronics, Communications and Networks, IEEE Publisher, 2012, pp. 1282−1285.
[29] L. Weimin, Z. Aiyun, L. Hongwei, Q. Menglin, W. Ruoqi, Dangerous goods dynamic monitoring and controlling system based on IOT and RFID, in: 24th Chinese Control and Decision Conference, IEEE Publisher, 2012, pp. 4171−4175.
[30] E.C. Anderson, D.O. Dike, Real time car parking system: a novel taxonomy for integrated vehicular computing, in: International Conference on Computing Networking and Informatics, IEEE Publisher, 2017, pp. 1−9.
[31] Y. Huang, C. Lung, Device with identity verification—apply in car driving as an example, in: IEEE International Conference on Applied System Invention, IEEE Publisher, 2018, pp. 243−246.
[32] J. Prinsloo, R. Malekian, Accurate vehicle location system using RFID, an Internet of Things approach, Sens. (Basel) 16 (825) (2016) 1−24.

CHAPTER

# AN EFFICIENT FRAMEWORK FOR OBJECT TRACKING IN VIDEO SURVEILLANCE

5

**D. Mohanapriya and K. Mahesh**

*Department of Computer Applications, Alagappa University, Karaikudi, India*

## CHAPTER OUTLINE



## 5.1 INTRODUCTION

Visual tracking is the process of effectively focusing the moving object over the human visual field. In image sequence, certain target is estimated by using visual tracking. A technique of generative tracking is an appearance model to represent their target, and image region is selected by using matching scores. Various challenges in tracking involve the computational complexity and increase their processing time. In visual tracking, there are many algorithms used, such as convolutional neural networks, hierarchical convolutional features, and similarity measuring, matching, and optimization techniques. Discriminative classification is used to classify the video images. Some tracking techniques are also used in the visual tracking systems, such as the foreground and background

The Cognitive Approach in Cloud Computing and Internet of Things Technologies for Surveillance Tracking Systems.
DOI: https://doi.org/10.1016/B978-0-12-816385-6.00005-2

tracker, Hough-based tracking [1], superpixel tracking, multiple-instance learning tracking, and tracking, learning, and detection. Visual tracking is utilized in many applications, such as:

- monitoring;
- surveillance and assistance;
- control and defense;
- robotics;
- autonomous car driving;
- human–computer interaction;
- mobile application and image stabilizing;
- action and activity recognition.

### 5.1.1 OBJECTIVES

The objectives of this work are as follows:

- to extract the foreground from the background;
- to track the objects that are in motion exactly;
- to associate target objects in consecutive video frames.

## 5.2 RELATED WORKS

Leal-Taixé et al. [2] present a benchmark for multiple-object tracking with the goal of creating a framework for the standardized evaluation of multiple-object tracking methods. It collects the two releases of the benchmark made and provides an in-depth analysis of almost 50 state-of-the-art trackers that were tested on over 11,000 frames.

Khalifa et al. [3] have proposed the detection of multiple objects using background subtraction methods and extract each object features by using the speed-up robust feature algorithm and then track the features through k-nearest neighbor processing from different surveillance videos sequentially. The background subtraction performed by subtracting the movement pixels from the static background objects.

In this paper, Danelljan et al. [4] describe the contribution of color in a tracking by the detection framework. The color attributes provide superior performance for visual tracking and an adaptive low-dimensional variant of color attributes. Both quantitative and attribute-based evaluations are performed with benchmark color sequences.

## 5.3 PROPOSED WORK

In this paper, we propose a novel model of background clustering by utilizing the neighborhood sequential-based pattern extraction (NSPE) algorithm for dissimilar background and object detection. Then, the pattern possibility analysis (PPA) technique is used for texture extraction in order to suppress the shadow pixels present in the frame. Then, patterns were extracted and classified using the machine learning classification (MLC), in which the moving objects were isolated from the nonmoving object.

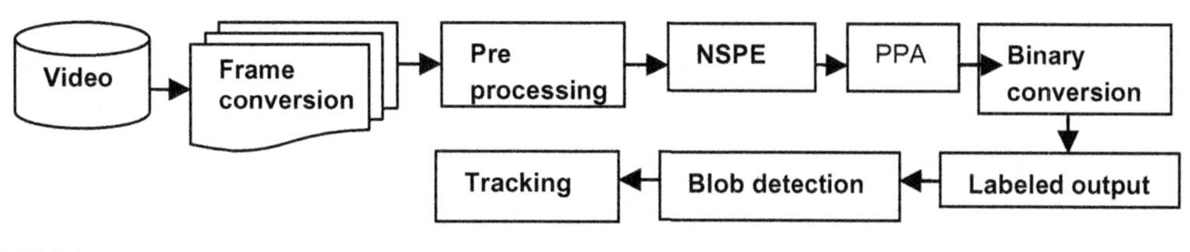


**FIGURE 5.1**

Overall flow of proposed work.

The schematic flow of the proposed system is represented in Fig. 5.1, where the image is preprocessed by the preprocessing block and is sent for the NSPE to transform the pattern of the video. The output of this is again processed by removing the shadow and is made to convert the video by utilizing the possibility analysis (PPA) technique that will also perform the weight of the pixel-dependent grid formation and the connected components. The proposed machine learning classification is provided by the training; afterwards, the retrieval of blob is carried out on the targeted tracking image.

## 5.4 PROPOSED PHASES

The major steps involved in this proposed work are as follows:

- preprocessing;
- object detection;
- feature extraction;
- object segmentation;
- object tracking.

### 5.4.1 PREPROCESSING

Preprocessing is the process of eliminating the noise and shadow segment in the video frames. Preprocessing is the primary step in video processing, in which the noise or the shadow region is detached using some filtering techniques. In this phase, the input frames are filtered using the median filter for the smoothening effect.

### 5.4.2 OBJECT DETECTION

The enhanced image from the preprocessing phase consists of interested objects as well as the background regions. The two main tasks of video tracking are object detection and recognition. Object detection, the process of searching each and every part of an image to localize the parts, is carried out, in which the properties of photometry or geometry should be matched with the targeted object in the training database. The similarity between the images is computed using correlation methods. The object detection is carried out to find the interested region using the NSPE technique to find object in each frame

### 5.4.3 FEATURE EXTRACTION

The feature extraction process is mainly used for reducing the quantity of the resources, which is required for describing a large set of data such as video sequences. While analyzing the complex large data, the major problems may raise due to the presence of a number of variables. This analysis of a large number of variables needs more amount of computation power and memory. Feature extraction is performed to extract the interested region PPA technique to extract object in each frame.

### 5.4.4 OBJECT SEGMENTATION

Object segmentation is the important task in the process of segmentation, which depends on the various features that are extracted from the image. Mostly, it consists of color or texture features. In the order of recovering the original image, the segmented features are denoising from an original image. The objective of the segmentation process is to decrease the information of an image for easy analysis. Also this segmentation helps to process the targeted regions more accurately.

### 5.4.5 OBJECT TRACKING

The most critical task in video processing is the tracking of moving objects as they move in a video. The issue of object tracking is to evaluate the locations and other associated information of the moving objects in an image classification. This process of object tracking is carried out for determining the activities of the object in the image frames.

The object or target is tracked by several methods and has drawbacks like merging with various frames, which has a lot of training features and intensity changes when suddenly changes happen and it is difficult to verify the matching features or points. To overcome these limitations of the existing methods, a novel or robust technique is proposed to track the target. The target or track region is represented with the bounding-box-tracked target region. The moving objects in the each frame are classified using the MLC algorithm, in which the classified objects are considered as the blobs. The blobs are the region of tracked object, and each frame and bounding box are applied for every blob in the video frame.

**ALGORITHM**

Step 1: Input video to frame conversion shown in Fig. 5.2
Step 2: Edge enhancement in converted frame
Step 3: Resize image to 256 × 256 (minimum Size)
Step 4: Edge Enhancement by applying Laplacian Equation
Step 5: Filter image using median filtering shown in Fig. 5.3
Step 6: Convert filtered RGB image to Gray Scale Image shown in Fig. 5.4
Step 7: Object is detected by using NSPE shown in Fig. 5.5
Step 8: Video sequence is tracked by PPA shown in Fig. 5.6
Step 9: Objects are classified by MLC shown in Fig. 5.7

**FIGURE 5.2**

Input video.

**FIGURE 5.3**

Preprocessing.

## 5.5 RESULTS AND DISCUSSIONS

The proposed work is implemented in windows 7 OS with Core-i5 and 3GB data set. In this research, there are two videos that are used to evaluate the proposed method. We have taken the Caviar data set and Walk-I video clips is done for experimental results.

**FIGURE 5.4**

Gray scale.

**FIGURE 5.5**

Blob detection.

**FIGURE 5.6**

Background subtraction.

**FIGURE 5.7**

Tracking.

### 5.5.1 ANALYSIS PARAMETERS

The proposed work is compared with the methodologies that are evaluated by true positive (TP), false positive (FP), true negative, and false negative (FN) computations.

#### 5.5.1.1 Precision

Precision metric is defined as the value assessed between the ratios of the associations between any two patterns retrieved at a single instance of time.

$$\text{Precision} = \frac{\text{TP}}{\text{TP} + \text{FP}}$$

The proposed work provides the precision value of 97.98% compared with the existing method hidden markov random function (HMRF) (64.2%).

#### 5.5.1.2 Recall

The proportional value inferred from those associated patterns and those retrieved patterns

$$\text{Recall} = \frac{\text{TP}}{\text{TP} + \text{FN}}$$

The proposed work provides a recall value of 95.7 compared with the existing method HMRF (64.8%).

#### 5.5.1.3 F-Measure(F)

F1-measure is defined as

$$F = 2 \cdot \frac{\text{Precision} \cdot \text{Recall}}{\text{Precision} + \text{Recall}}$$

#### 5.5.1.4 Success and failure rate

| Tracking Methods | Success Rate | Failure Rate |
| --- | --- | --- |
| PartT | 54.93 | 23.3 |
| MVS | 90.14 | 6 |
| NSPE-PPA | 98.95 | 4.86 |

MVS, Multiview SVM (SVM-Support vector model); NSPE-PPA, *Neighborhood sequential-based pattern extraction-pattern possibility analysis.*

## 5.6 CONCLUSION

In this novel method for foreground region analysis, the method of NSPE and the possibility analysis (PPA) are introduced for detecting the movable objects and extracting the features by suppressing the foreground and background regions. This is because of tracking the targeted region from the input image. The input image is retrieved and is preprocessed via some techniques, and the performance of NSPE-PPA is attained for the conversion of video on utilizing the formation of pixel grids along with the connected component extraction. The image is then tracked on using the machine learning classification algorithm to attain the targeted output image pattern. This section also provides the estimation of performance on measuring the accuracy, precision, recall, F-measure, success rate, and failure rate of the nearest neighbor chain prediction and the differential boundary pattern. The performance analysis is made for proving the efficiency and improvement of this proposed method to that of the existing methodologies. Hence, the proposed method shows the effective improvement of accuracy, success rate, and so on with the reduction in the rate of error.

## ACKNOWLEDGMENT

I thank EC Funded CAVIAR project/IST 2001 37540 found at URL: http://homepages.inf.ed.ac.uk/rbf/CAVIAR/ for using data for my research work and also I would like to express my special thanks to Dr. V. Palanisamy, Professor, Department of Computer Applications, Alagappa University, for giving full support and DST-PURSE, which help me in doing research work.

## REFERENCES

[1] D. Anguita, A. Ghio, L. Oneto, X. Parra, J.L. Reyes-Ortiz, 2013. A public domain dataset for human activity recognition using smartphones, European Symposium on Artificial Neural Networks, Computational Intelligence and Machine Learning.

[2] L. Leal-Taixé, A. Milan, K. Schindler, D. Cremers, I. Reid, S. Roth, Tracking the trackers: an analysis of the state of the art in multiple object tracking, arXiv preprint arXiv:1704.02781, 2017.

[3] O. Khalifa, N.A. Malek, K.I. Ahmed, Robust vision-based multiple moving object detection and tracking from video sequences, Indones. J. Electr. Eng. Comput. Sci. 10 (2) (2018) 817–826.

[4] M. Danelljan, F. Shahbaz Khan, M. Felsberg, J. Van de Weijer, Adaptive color attributes for real-time visual tracking, in: IEEE Conference on Computer Vision and Pattern Recognition (CVPR), Columbus, OH, June 24–27, 2014, pp. 1090–1097.

## FURTHER READING

C. Garate, P. Bilinsky, F. Bremond, Crowd event recognition using hog tracker, in: Performance Evaluation of Tracking and Surveillance (PETS-Winter), 2009 Twelfth IEEE International Workshop, 2009, pp. 1–6.

M. Kaâniche, Gesture recognition from video sequences, degree, Université Nice Sophia Antipolis, Nice, France, 2009.

T. Lan, L. Sigal, G. Mori, Social roles in hierarchical models for human activity recognition, in: Computer Vision and Pattern Recognition (CVPR), 2012 IEEE Conference, 2012, pp. 1354–1361.

W. Liu, T. Mei, Y. Zhang, C. Che, J. Luo, Multi-task deep visual-semantic embedding for video thumbnail selection, in: Proceedings of the IEEE Conference on Computer Vision and Pattern Recognition, 2015, pp. 3707–3715.

D. Mohanapriya, K. Mahesh, A novel foreground region analysis using NCP-DBP teture pattern for robust visual tracking, Multimed. Tools Appl. 76 (24) (2017) 25731–25748.

G. Ning, Z. Zhang, C. Huang, X. Ren, H. Wang, C. Cai, et al., Spatially supervised recurrent convolutional neural networks for visual object tracking, in: Circuits and Systems (ISCAS), 2017 IEEE International Symposium, 2017, pp. 1–4.

R.K. Rout, A survey on object detection and tracking algorithms, degree, 2013, pp. 1–75.

S. Zhang, X. Yu, Y. Sui, S. Zhao, L. Zhang, Object tracking with multi-view support vector machines, IEEE Trans. Multimed. 17 (3) (2015) 265–278.

CHAPTER

# 6 DEVELOPMENT OF EFFICIENT SWARM INTELLIGENCE ALGORITHM FOR SIMULATING TWO-DIMENSIONAL ORTHOMOSAIC FOR TERRAIN MAPPING USING COOPERATIVE UNMANNED AERIAL VEHICLES

**G. Pradeep Kumar[1] and B. Sridevi[2]**

[1]*Velammal College of Engineering and Technology, Madurai, India*

[2]*Velammal Institute of Technology, Chennai, India*

## CHAPTER OUTLINE



The Cognitive Approach in Cloud Computing and Internet of Things Technologies for Surveillance Tracking Systems.

DOI: https://doi.org/10.1016/B978-0-12-816385-6.00006-4

## 6.1 INTRODUCTION

An unmanned aerial vehicle (UAV), commonly known as a drone, is an aircraft or an airborne system that is remotely operated by a human operator or autonomously by an onboard computer. An UAV can be defined as a powered, aerial vehicle that does not carry a human operator but works autonomously by an on board computer. They are used for a variety of applications, especially in military applications. UAVs are increasingly used, because they have the advantage of not placing human life at risk and lowering operational costs. To realize these advantages, UAVs must have a higher degree of autonomy and preferably work cooperatively in groups. Groups of UAVs that work together as a single unit to complete a particular mission, such as mapping, are referred to as cooperative UAVs.

Our project focuses on the simulation of cooperative UAVs with embedded path planning and collision avoidance algorithms, such that they work autonomously and capture the images required for two-dimensional (2D) orthomosaics. An orthomosaic is a 2D map. It is created by using a series of aerial images of a particular area and by stitching the overlapping areas together. Thus a group of UAVs working cooperatively can get the images faster and more accurately than a single UAV. In this chapter, we have simulated the UAVs using the DroneKit software in the loop (SITL) and Mission Planner. The advantage of our method is that we can easily debug and modify the design, as it is in simulation. In addition, we can easily embed this in hardware, as the software emulates that in the hardware board.

## 6.2 LITERATURE REVIEW

### 6.2.1 EFFICIENT THREE-DIMENSIONAL PLACEMENT OF A UNMANNED AERIAL VEHICLE USING PARTICLE SWARM OPTIMIZATION

When cellular networks exhaust, UAVs can be utilized as aerial wireless base stations. In the literature, UAV-based wireless coverage is considered as an air-to-ground path loss model, which is based on the assumption that the users are outdoor and they are positioned on a 2D plane. In this chapter, a single UAV is designed to provide wireless coverage for the users inside a high-rise building under disaster conditions (such as earthquakes or floods). It is assumed that the locations of indoor users are uniformly distributed in all the floors, and a particle swarm optimization (PSO) algorithm is proposed to find the efficient three-dimensional (3D) positioning of a UAV that minimizes the total transmit power required to cover the indoor users. At a low altitude, the data of path loss between the aerial vehicle and the ground user reduce, when the probability for the line of sight links decreases. Alternatively, at a high altitude, the line of sight connections exist with a high probability, while the path loss increases. Anyhow, it is assumed that all users are outdoor and the location point of each user can be represented by an outdoor 2D point

### 6.2.2 API DEVELOPMENT FOR COOPERATIVE AIRBORNE-BASED SENSE AND AVOID IN UNMANNED AIRCRAFT SYSTEM

The usage of unmanned aerial systems (UASs) to civil applications has extended the research interest in this field. Under a common airspace, various types of Unmanned Aerial Systems are used for detection of target. But the applications are limited due to various reasons. This is because of the complexity of the surrounding environment, the nonsegregated airspace, the variety in UASs, both from software and hardware prospective, and the wide use of UASs in civil applications. The purpose of this chapter is to extend the functionality of a famous open-source software development kit (SDK) for an open-platform UAS, DroneKit, in the advanced control spectrum. The infrastructure for a decentralized solution for cooperative conflict detection has been developed, containing support for inter-UAS communication, prioritization, and mission save/restore in a resource conscious and developer-friendly application program interface (API). The software developed undergoes both laboratory and field testing for evaluation purposes. The results are discussed, and the advantages of the system, along with its limitations, are presented.

### 6.2.3 MULTIPLE-SCENARIO UNMANNED AERIAL SYSTEM CONTROL: A SYSTEMS ENGINEERING APPROACH AND REVIEW OF EXISTING CONTROL METHODS

The use of UASs in both the public and military environments is expected to grow significantly. The demand for UASs grows, and then the availability of more robust and capable vehicles that can perform multiple mission types will be required. In the public sector, the demand will grow for UASs for agriculture, forestry, and search and rescue missions. Militaries have demand over UAS capabilities for diverse operations around the world. A significant research has been performed and continues to progress in the areas of autonomous UAS control. Most of the work focuses on the subsets of UAS control: path planning, autonomy, small UAS controls, and sensors. The short work exists on a system-level issue of multiple-scenario unmanned autonomous system control for integrated systems. This chapter gives a high-level modular system architecture definition that is modifiable across platform types and mission requirements. A review of the current research and employment of UAS capabilities is provided to evaluate the states of the capabilities required to enable the proposed architecture. The systems being developed will need to be safe, reliable, and effective across multiple operational environments and tasks, be able to perform under multiple scenarios, and be adaptable to new capabilities and responsive to changes in environments and missions, which will be key to future success.

### 6.2.4 FLOCKING ALGORITHM FOR AUTONOMOUS FLYING ROBOTS

The beast swarms displaying different typical flocking patterns would not exist without satisfying safe, optimal, and stable dynamics of any individual. These patterns can be efficiently reconstructed with simple flocking models, based on three simple rules: cohesion of the flock, repulsion of neighboring individuals, and alignment of velocity between neighbors. When designing robust swarms, the controlling adoptive dynamics of the robots can be based on the models. In this chapter, we present such a flocking algorithm, endowing flying robots with the capability of self-organized collective maneuvering. The main feature of the work is that we include the velocity alignment part of the equations, which is an analogue of the usual frictional force between point-wise bodies. We introduce a generalized mathematical model of an autonomous flying robot, based on flight field tests. Using simulations, we test the flocking algorithm from the aspects of the most general deficiencies of robotic systems such as time delay, locality of the communication, and inaccuracy of the sensors. Some of these deficiencies often cause instabilities and oscillations in the system. We show that the instabilities can be efficiently reduced in all states of the system by the inclusion of the friction-like velocity alignment, resulting in stable flocking flight of the robots. By using a realistic simulation framework and studying the group behavior of autonomous robots, we can learn about the majority of factors influencing the flight of bird flocks.

### 6.2.5 A GROUND CONTROL STATION FOR A MULTIUNMANNED AERIAL VEHICLE SURVEILLANCE SYSTEM

This proposed work presents the ground control station (GCS) developed for multiple UAVs in surveillance missions. The application is founded on open-source libraries, and it has been intended as a robust and decentralized system. It allows the user to dynamically allocate variety of jobs to the vehicles and to show their operational information in a real-time 3D environment. The ground station is to assist the user in the different challenging tasks of controlling a system with multiple vehicles, initiating to reduce the workload. The multi-UAV cooperative surveillance system has been shown in field experiments with two quadcopters equipped with cameras. The design adopted on the system architecture has demonstrated to be an efficient solution to achieve a rapid deployment of a multi-UAV surveillance system. The design of the GCS exploited the autonomous capabilities of the multi-UAV system to decrease the workload of the operator. Thus the developed GCS allowed exploiting most of the capabilities that the autonomous multi-UAV system could provide, while at the same time offered a user-friendly and easy-to-operate interface. The integration of the information from other sensors in the GCS (not only ground cameras) can be accessed from a remote location through the internet has been done.

### 6.2.6 MULTIUNMANNED AERIAL VEHICLE CONTROL WITH THE PAPARAZZI SYSTEM

Two experiments involving multi-UAV control are presented. At first, three vehicles flew in formation flight, controlled by a ground station. Then, two aircraft flying in different places in the GPS have been controlled from the same GCS from somewhere. Both experiments have been conducted continuously in the Free Autopilot. The human interactions with system were complex during those experiments. Controlling three aircrafts in the same time in a safe way requires the help of a carefully studied interface. Controlling a UAV from a distant remote place requires sharing the control authority between the local and distant operators. We believe that the human work for UAV's control and command in this context is more similar to the work of an air-traffic controller than to the work of a pilot. The graphical interface of the system is developed with the need. These tests involve multi-UAV control by several operators in the same time. The results show that the design of Paparazzi provides the perfect architecture for distributed agents, UAVs, or/and operators. The time in hours of flight shows that the system must be autonomous and enough to provide time to the operator to react to unexpected events. The aerial vehicle or microaerial vehicle must autonomously stay in the air environment as long as possible, and audible or visible alarms are shown to the operator. Paparazzi is an open-source autopilot system, oriented toward inexpensive autonomous aircrafts of all types. The more number of autonomous flights have been successfully achieved with the Paparazzi system.

## 6.3 RELATED WORKS

The simulation work depends on the following process for implementing the efficient cooperative UAVs.

### 6.3.1 COOPERATIVE UNMANNED AERIAL VEHICLE METHODS

With the advancement in advanced sensing and information technology, cooperative UAV control can be achieved by using a variety of sensors. The existing methods deal about controlling the multiple UAVs using one remote controller. The safe flight of multiple UAVs can be done by maintaining a particular distance between each UAV. The distance maintenance behavior can be achieved by using the principle that if two copters are so close to each other, then the copters must move away autonomously. Similarly, if the copters are away from each other while locating the position, then the copters must move closer autonomously. UAVs send their GPS coordinates to other UAVs, thereby maintaining the distance. Cooperative UAV control can be achieved by using sensors and remote controllers.

### 6.3.2 PATH PLANNING

The key factor for cooperative UAVs is path planning. The existing methods for path planning can be classified into two types: (1) predefined flight path-based search; and (2) dynamic path planning. In the predefined flight path method, first the flight paths are generated in advance, and the paths are followed during the execution. They are done by using a sweep line-based search. This method is so effective; hence, no search areas are missed but not efficient methodologies due to the predefined paths. In addition, this method cannot be used for searching the dynamic targets. Dynamic programming, artificial intelligence, and model predictive control can be used for efficient path planning. The optimal path has to be found out. There are several algorithms for path planning, and some of them are sampling-based algorithms, node-based algorithms, mathematical model-based algorithms, bioinspired algorithms, and multi-fusion-based algorithms.

### 6.3.3 COLLISION AVOIDANCE

The UAVs are required to fly in the defined path and to avoid collisions between each other. In the older methods, the collision avoidance between the vehicles is achieved by using a dual mode control system, namely safe and danger mode. The safe mode is activated when there is no obstacle in the free environment, whereas the danger mode is defined when there are chances for collisions in the path. Every collision avoidance system contains two vital parts, and they are sensing and detection. The sensing process defines about getting vital information about the environment around the vehicle. Here, the collision avoidance generally refers to the vehicle that can acknowledge dangers that are not been initially known and act concurrently. For sensing some sensors, like ADS-B, visual sensor or radar can be used. Once an UAV senses an obstacle, it has to determine whether it will cause collision or not. There are a few collision detection methods, such as trajectory calculation and distance estimation, worst case, probabilistic, and act as seen.

## 6.4 PROPOSED ARCHITECTURE

Here, the DroneKit software is used for the simulation of the UAV. The DroneKit SITL is used for simulating the vehicle without the real one. We need virtual machine for simulating multiple drones as the SITL can simulate only one UAV. DroneKit allows us to control the UAV using Python programming language and to test bug fixes and other changes to the autopilot. This method uses MAVproxy to make the initial connections. As shown in Fig. 6.1, we run multiple machines in the virtualbox and in each machine we run the DroneKit SITL. They each connect to Mission Planner using either a transmission control protocol (TCP) or a user datagram protocol (UDP) connection. Mission Planner provides a virtual environment to simulate the vehicle and implement the path planning and collision avoidance algorithms. We can set the waypoints and targets in Mission Planner.

### 6.4.1 DRONEKIT-PYTHON

DroneKit-Python is an open-source and community-driven project. It is installed by using a Python pip tool on all platforms. It is a project of ArduPilot, created for connecting, controlling, and monitoring a vehicle. DroneKit helps you to create powerful applications for UAVs. These applications

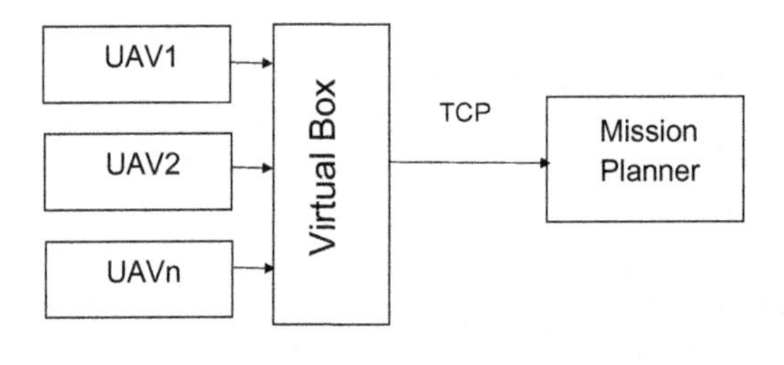


**FIGURE 6.1**

Block diagram.

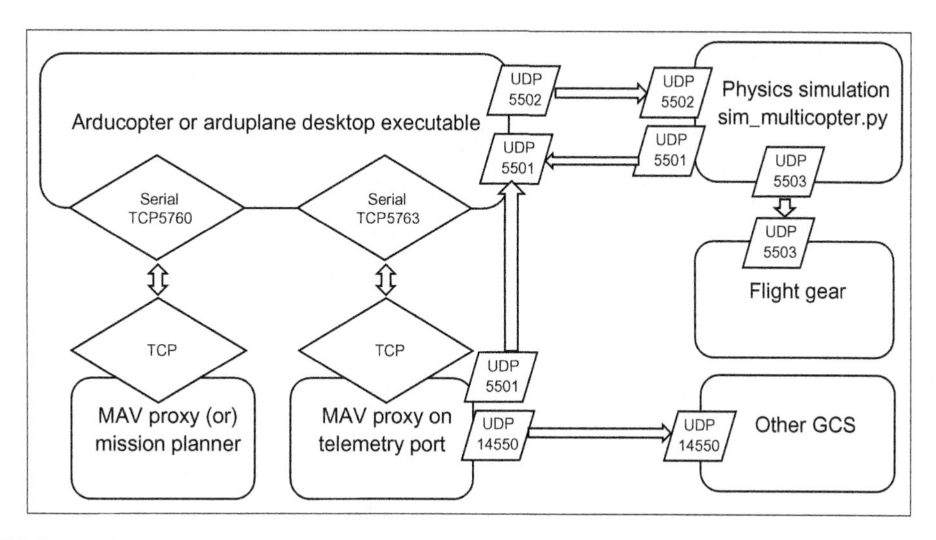


**FIGURE 6.2**

Block diagram of ArduPilot modules.

can run in their companion computers. They can also perform tasks that are computationally intensive and use a low latency link. It provides compatibility with vehicles that communicate using the microair vehicle link (MAVLink) protocol. One of its major advantages is that it uses Python programming language (Fig. 6.2).

Python is a high-level programming language and easy to program and interpret. DroneKit-Python operates on Linux, Mac OS, or Windows.

#### *6.4.1.1 Installation*

DroneKit-Python can be installed by the following command:

```
pip install dronekit
```

### 6.4.2 DRONEKIT-PYTHON SOFTWARE IN THE LOOP

DroneKit SITL is the fastest, simplest, and easiest way to simulation on Windows, Linux, or Mac OS X.

The SITL simulator can be used to test the algorithms and processes for UAVs without the real/physical vehicle.

#### *6.4.2.1 Installation*

DroneKit SITL can be installed in all platforms using the following command:

```
pip install dronekit-sitl
```

#### 6.4.2.2 Running software in the loop

We can run any SITL vehicles using the default settings

```
dronekit-sitl copter
```

or can input parameters such as home location, vehicle model type, (e.g., quad), etc.

```
dronekit-sitl plane-3.3.0 --home=-35.363261,149.165230,584,353
```

### 6.4.3 MAVLINK

MAVLink is a protocol for communicating with unmanned vehicles. It is used for communication between a GCS and unmanned vehicles, and in the intercommunication of the subsystem of the vehicle. It can also be used to transmit the state of vehicle, its GPS coordinates, and other parameters.

### 6.4.4 ARDUPILOT

ArduPilot is an open-source platform, which is the highly advanced application, capable of controlling the vehicle system designable, from conventional airplanes, submarines, hexacopters, helicopters, to even boats. It has both hardware and software, thus allowing for testing before implementation.

### 6.4.5 MISSION PLANNER

Mission Planner is a full-featured ground station application for the ArduPilot open-source autopilot project. It has various options and features that help in monitoring and controlling the vehicle. In Mission Planner, we can view the location of the vehicle, its state parameters, roll, pitch, and yaw (RPY) axis, and plenty of other options. We can also record the running of the vehicle as a video. Waypoints and missions can be programmed and can even be saved for later use. We can view the running of the DroneKit SITL in Mission Planner using a TCP or UDP connection.

### 6.4.6 TWO-DIMENSIONAL ORTHOMOSAICS

An aerial shot image data set along with a geometrical reference has been corrected orthophoto, orthophotograph, or orthoimage, such that the scale is uniform. The photo has the same dearth of distortion in the map. A 2D orthomosaic is basically a 2D map created from a series of aerial images taken by the UAVs. The drones work as a unit and capture the images of a particular area from various angles.

These images can be used to form a large image using an algorithm that stiches together the images using color detection. A single UAV can also be used to create an orthomosaic image, but the efficiency and the accuracy increase by using cooperative UAVs.

## 6.5 SIMULATION OF THE DRONEKIT SOFTWARE IN THE LOOP

After installing DroneKit and the DroneKit SITL, we can run the SITL for various vehicles such as copter, plane, quad, etc. To run the latest version of a copter, we can simply call:

```
dronekit-sitl copter
```

The SITL will trigger and wait for TCP connections on 127.0.0.1:5760, where 5760 is the port number. We can also specify the version and other parameters of the vehicle.

We can connect Mission Planner to the DroneKit SITL after running it (Fig. 6.3). In addition, we can run the other.

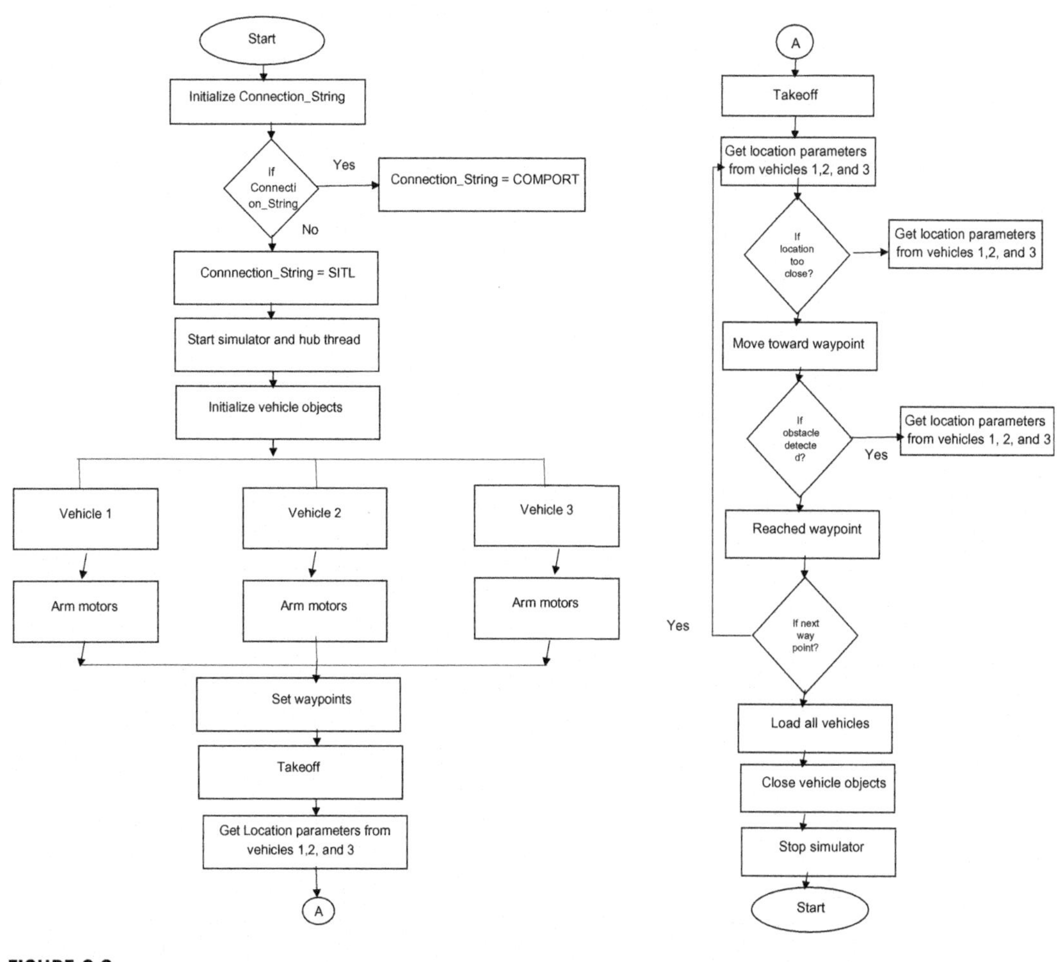


**FIGURE 6.3**

Flow of execution.

**FIGURE 6.4**

Parameter selection in Mission Planner.

The initial calibration of each and every vehicle can be configured by using this software in terms of RPY. In case of octocopters, all the wings and propellers should be configured in order to maintain the stability (Fig. 6.4).

Python programs in the command prompt view the results in Mission Planner. The graph, as shown in Fig. 6.5, of Mission Planner is the graph of the RPY axis while the drone is flying. The channel for communication with each other can also be configured in the software.

Initially, single UAV has configured and flied along with the GPS location to study the parameter outcome for the efficiency calculation.

For multiple UAVs, we can use Oracle VirtualBox. In the virtualbox, we can run the DroneKit SITL in different virtual machines. In Fig. 6.6, we are running two Ubuntu machines, and in each machine, we run ArduCopter.elf (Figs. 6.7–6.9).

We can then connect Mission Planner to both vehicles using a TCP connection.

In Mission Planner, we can create the waypoints for the UAV to fly (Fig. 6.10).

According to the GPS data, the vehicle will start moving along with the velocity assigned in the algorithm. In the traditional PSO algorithm, the velocity gets changed according to the real-time data of wind, energy consumption, and so on. In the modified PSO, the velocity keeps constant. While the vehicle meets the wind at the real time, the rpm of propellers will get vary to maintain the stability and the velocity. Here, the velocity becomes constant. This is the modification that has been done in the algorithm that provides low probability of collision occurrence when compared with the traditional PSO algorithm. Fig. 6.11 shows the simulation of multiple UAVs flying together by communicating with each other according to the GPS data. This proposed algorithm can be used for the real-time terrain mapping of the largest area. The vehicles split the mapping work and stitched the pieces into one large area by the process of 2D orthomosaic (Table 6.1).

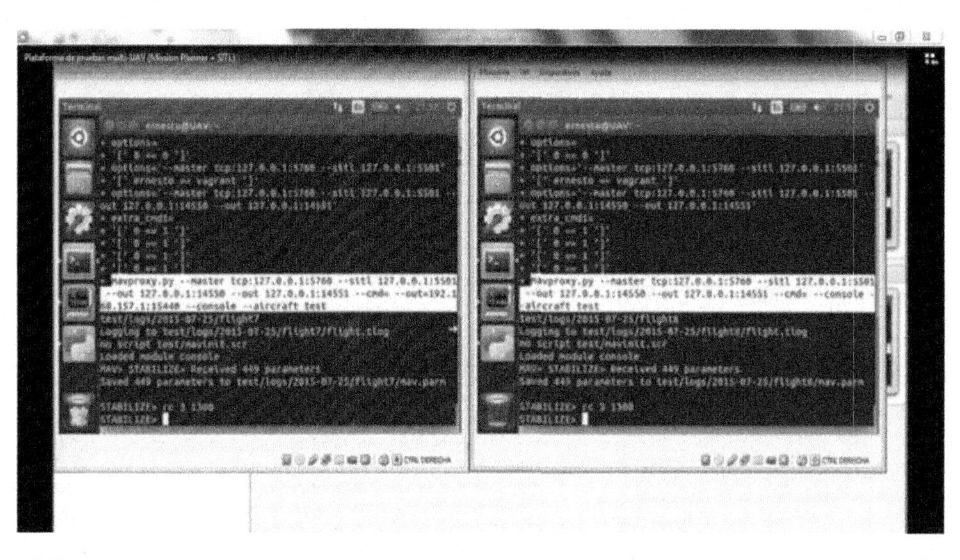

**FIGURE 6.5**

DroneKit SITL running in two virtual machines.

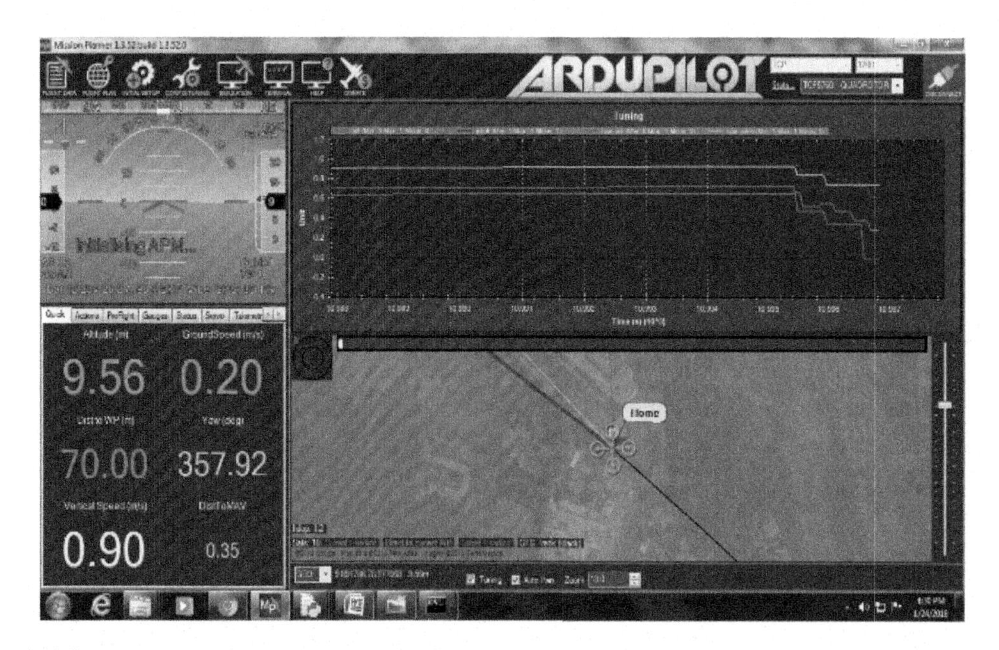


**FIGURE 6.6**

Quad flying over a given location.

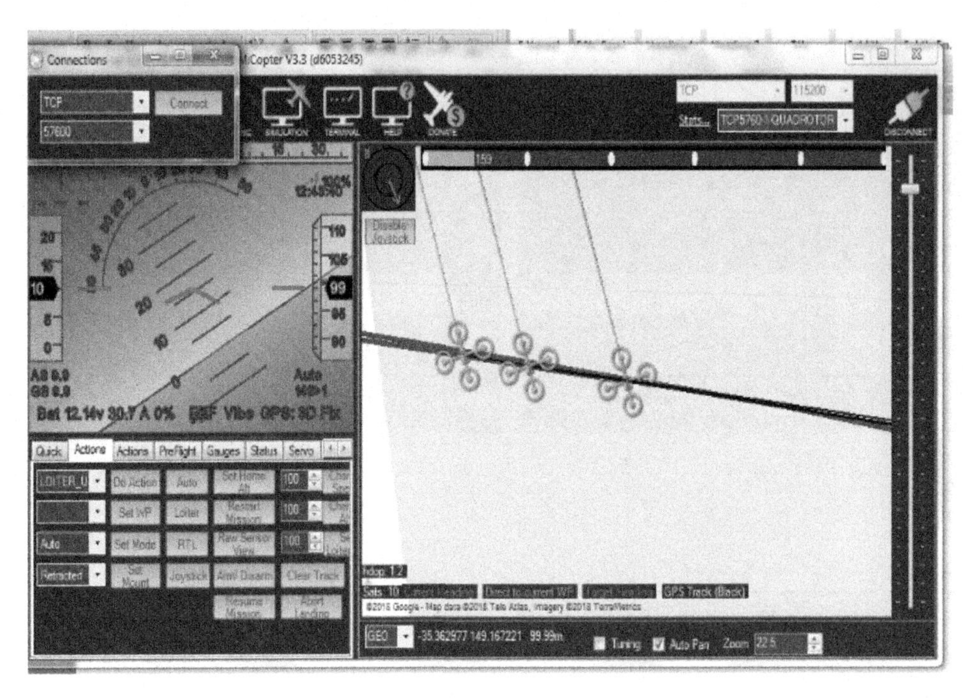


**FIGURE 6.7**

Simulation of takeoff of three UAVs.

**FIGURE 6.8**

Simulation of UAVs in Mission Planner.

**FIGURE 6.9**

Setting the global positioning system location to initiate.

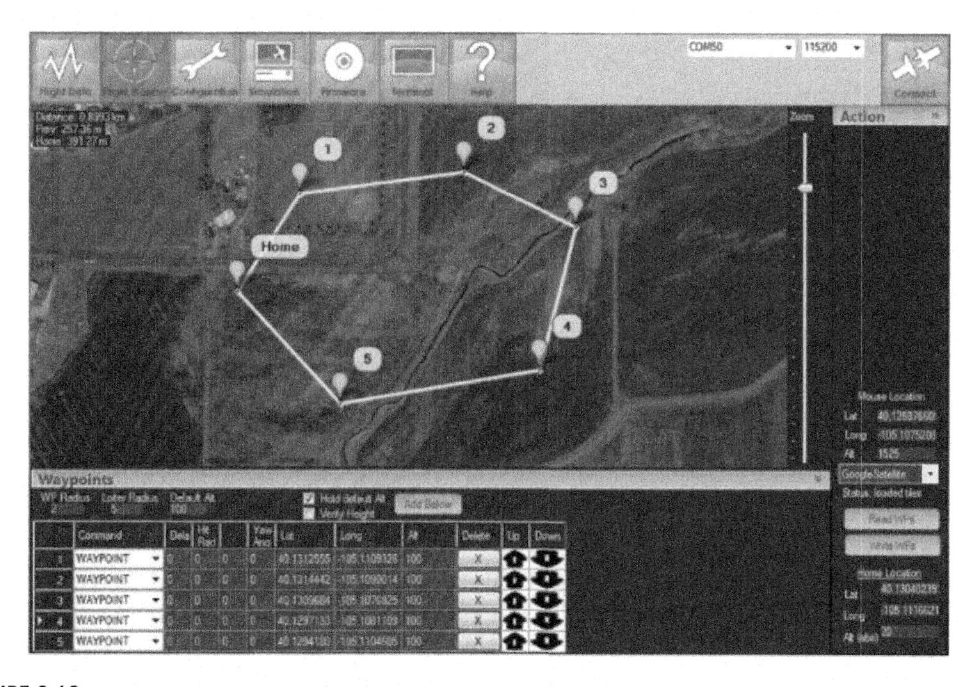


**FIGURE 6.10**

Simulation of multiple UAVs.

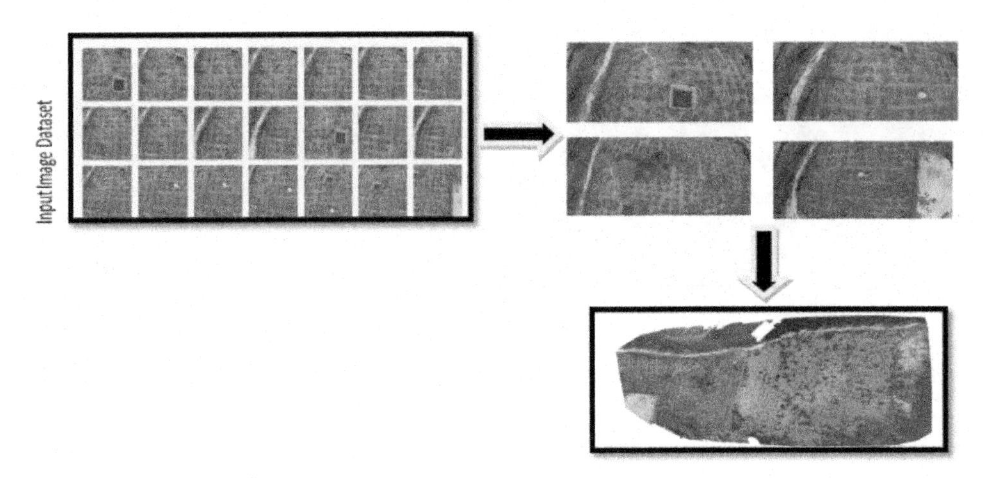


FIGURE 6.11

Processing of images in two-dimensional orthomosaics.

**Table 6.1 Readings From Mission Planner**

| Parameters | UAVs | | |
|---|---|---|---|
| | UAV 1 | UAV 2 | UAV 3 |
| Altitude (m) | 1.20 | 0.97 | 2.38 |
| Battery remaining (%) | 92 | 99 | 94 |
| Ground speed (m/s) | 0.37 | 0.24 | 0.14 |
| Vertical speed (m/s) | 6 | 5 | 4 |
| Distance to (Wind pressure)WP (m) | 0.00 | 0.08 | 0.06 |
| Distance traveled (m) | 3.5 | 3.2 | 4.1 |
| Yaw (degrees) | 154.17 | 344.36 | 99.19 |
| Wind speed (m/s) | 3 | 1.8 | 2.2 |

## 6.6 COLLISION AVOIDANCE AND PATH PLANNING

In cooperative UAVs, the major challenges are obstacle/collision avoidance and path planning. While the groups of UAVs are flying, they need to discover a path that is obstacle free and does not cause them to collide with each other. To achieve this, we need an algorithm to find the best possible path while the UAV is flying. The various algorithms that can be used to find the best path are Dijkstra's algorithm, Bellman Ford's algorithm, Floyd–Warshall's algorithm, and the AStar algorithm. Among these algorithms, the AStar algorithm is found to be the best, as it does not choose the next state only with the lowest heuristic value but the one that gives the lowest value when considering its heuristics and cost of getting to that state.

**Table 6.2 Comparison of PSO and Modified PSO**

| | PSO | | | Modified PSO | | |
|---|---|---|---|---|---|---|
| **Parameters** | **Vehicle 1** | **Vehicle 2** | **Vehicle 3** | **Vehicle 1** | **Vehicle 2** | **Vehicle 3** |
| Velocity (m/s) | 6.1 | 5.033 | 4.465 | 6.03 | 6.029 | 6.032 |
| Fuel efficiency (%) | 64 | 72 | 71 | 76 | 81 | 79 |
| Average cost time (ms) | 76 | 80 | 93 | 65 | 67 | 73 |

PSO, *Particle swarm optimization.*

**Table 6.3 Graph of Velocity.**

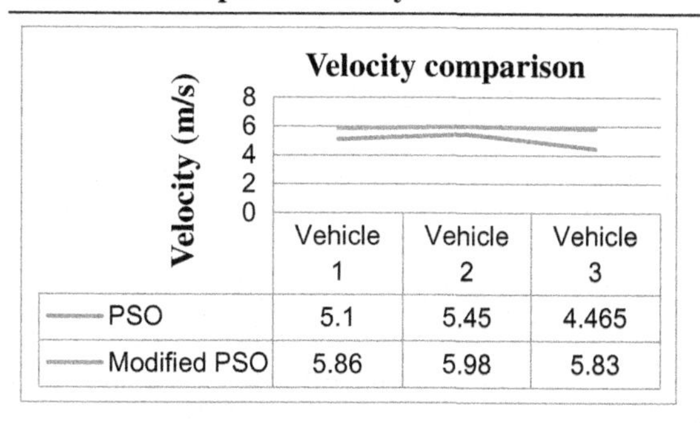


The collision avoidance is made possible by getting the GPS location of each drone. If the drones move too close to the critical distance of each other, then they are made to move farther. If they are too far away, then they are made to come closer to maintain the formation. Thus, by this method, we achieve both collision avoidance and path planning.

The above data set has been taken for three vehicles flown cooperatively with a constant velocity. While having the constant velocity, the collision occurrence gets reduced. The altitude also fixed for all the vehicles around the same integer. The algorithm plays a vital role for maintaining the constant velocity, but the battery drains easily because of the sudden changes of rpm in the propellers (Tables 6.2 and 6.3).

Here, in the proposed work, we focus only on the velocity.

## 6.7 APPLICATIONS

Orthomosaic maps are a group of many overlapping images of a defined area, which are processed to create a new, larger "orthomosaic": a highly detailed, up-to-date map that is in true scale. The orthomosaic has a uniform scale and can be used for 2D measurements, distances, and areas, because it is an accurate representation of the Earth's geographic surface. The images are taken by the camera fixed in the UAVs. These images are then processed by the orthomosaic algorithm. The

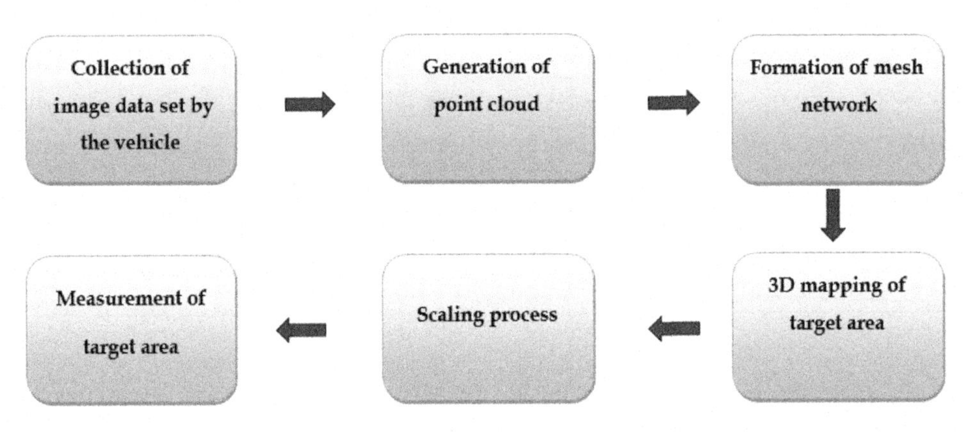


**FIGURE 6.12**

Block diagram of three-dimensional mapping.

**FIGURE 6.13**

Measurement of three-dimensional object.

algorithm stiches the photos together using color detection. Thus we can obtain the complete map using this method. The advantage of orthomosaics is that distances are preserved and therefore the map can be used for measurements.

As further development, in the real-time implementation in hardware, 3D modeling of the whole area can be generated by using the 2D image data set generated by the vehicle. The image data set will be fed to the modeling software as input, and based on the input images, the point cloud generation and formation of mesh network using the points will create the complete 3D modeling using software like Meshlab, Photomodel, and so on (Figs. 6.12–6.15).

This process of 3D modeling is used in many applications: landslide prediction, mine survey, terrain mapping, measurement of historical buildings and objects, and underwater archeological survey.

The reconstruction process might be very useful in the archaelogical process, especially the underwter archaelogical process. Some objects can be buried in the water. Those kinds of objects can be taken as 2D images, and the complete object can be rebuilt by using this process. The proposd work initiated the cooperative UAV flying for terrain mapping and the 3D modeling of any geographical area.

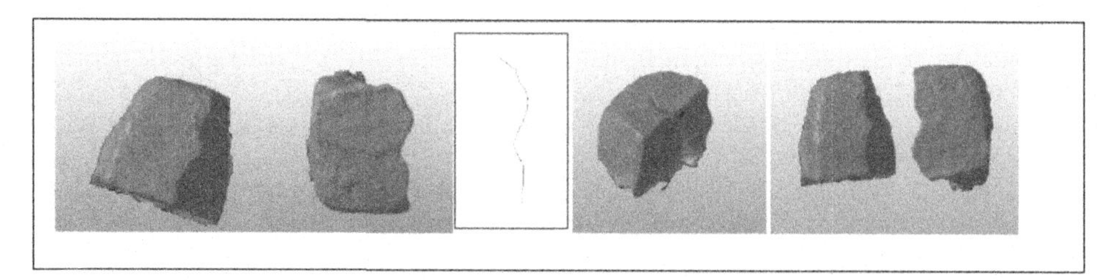

**FIGURE 6.14**

Reconstruction of an object in three-dimensional modeling.

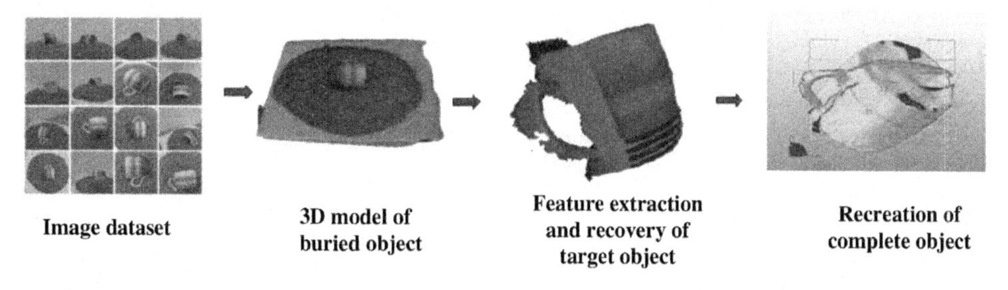


**FIGURE 6.15**

Reconstruction of an object.

## 6.8 CONCLUSION

In this chapter, we have simulated the cooperative UAV with the collision avoidance and path planning algorithm, shown the simulation of terrain mapping using cooperative UAVs. The image data set has been taken from the software while simulating along with the GPS location. This efficient application can be implemented in real time by having the control board of Raspberry PI for image processing and Pixhawk for controlling the vehicle without collision. The same proposed work can also be used for precision agriculture, 3D mapping, emergency rescue missions, and various other applications.

## FURTHER READING

[1] C.M. Eaton, E.K.P. Chong, A.A. Maciejewski, Multiple-Scenario Unmanned Aerial System Control, January 2016, Aerospace 2016, https://doi.org/10.3390/aerospace3010001

[2] X. Wang, V. Yadav, S.N. Balakrishnan, Cooperative UAV formation flying with obstacle/collision avoidance, IEEE Transactions on Control Systems Technology 15 (4) (July 2007).

[3] S. Lazzaro, Flying Multiple Drones From 1 Remote Controller, April 2008, Own paper supervised by Michael Coen.

[4] C. Virágh, G. Vásárhelyi, N. Tarcai, T. Szörényi, G.ő. Somorjai, T. Nepusz, et al., Flocking algorithm for autonomous flying robots, Bioinspir. Biomim. 9 (2) (October 2013). Available from: https://doi.org/10.1088/1748-3182.

[5] C. Trouwborst, Control of Quadcopters For Collaborative Interaction, CTIT, University of Twente, Report nr. 006RAM2014, July 2014.

[6] X. Ji, X. Wang, Y. Niu, L. Shen, Cooperative search by multiple unmanned aerial vehicles in a nonconvex environment, Robot. Mecahtron 2015 (July 2014). Article ID 196730.

[7] Y. Ben-Asher, M. Feldman, S. Feldman, P. Gurfil, Distributed decision and control for cooperative UAVs using ad-hoc communication, IEEE Transactions on Control Systems Technology 16 (3) (May 2008).

[8] A. Zulu, S. John, A review of control algorithms for autonomous quadrotors. Open J. Appl. Sci. 4 (2014) 547–556.

[9] B.S.E.E. Prateek Burman, Quadcopter Stabilization with Neural Network, University of Texas at Austin, December, 2016. Report supervised by Christine Juline 2016.

[10] H. de Plinval, A. Eudes, P. Morin, Control and estimation algorithms for the stabilization of VTOL UAVs from mono-camera measurements, Aerospace lab J (June 17, 2014). Available from: https://doi.org/10.12762/2014.AL08-07.

[11] C. Roberts, Electrical Engineering "GPS Guided Autonomous Drone," University of Evansville, Indiana, Supervised by Dr.Tony Richardson, April 25, 2016.

[12] T. Ryan, H.J. Kim, Smartphone quadrotor flight controllers and algorithms for autonomous flight, Renewable Energy Program of the Korea Institute of Energy Technology Evaluation and Planning (KETEP) December 2017.

[13] L.E. Romero, D.F. Pozo, J.A. Rosales, Quadcopter Stabilization by Using PID Controllers, Maskana, I + D + ingenieria , IEEE Transaction, 2014.

[14] S.A. Quintero, Optimal UAV coordination for target tracking using dynamic programming, in: Decision and Control (CDC), IEE Transaction, https://doi.org/10.1109/CDC.2010.5717933 ISBN 9781424477456, 2010.

[15] Y. Ben-Asher, M. Feldman, S. Feldman, P. Gurfil, Distributed Decision and Control for Cooperative UAVs Using Ad-Hoc Communication, IEEE Transactions on Control Systems Technology 16 (3) (May 2008).

[16] H. Shakhatreh, A. Khreishah, A. Alsarhan, I. Khalil, A. Sawalmeh, N.S. Othman, Efficient 3D placement of UAV using particle swarm optimization, 2017 8th International Conference on Information and Communication Systems (ICICS), https://doi.org/10.1109/IACS.2017.7921981

[17] Alan Vande Wouer, Cooperative search by multiple UAV in a nonconvex environment, in Mathematical Problems in Engineering, Volume 2015, Article ID 196730

[18] Z. Md. Fadlullah, D. Takaishi, H. Nishiyama, N. Kato, R. Miura, A dynamic trajectory control algorithm for improving the communication throughput and delay in UAV-aided networks, IEEE Netw. 30 (1) (2016) 100–105.

[19] V.A. Maistrenko, L.V. Alexey, V.A. Danil, Experimental estimate of using the ant colony optimization algorithm to solve the routing problem in FANET, in: 2016 International Siberian Conference on Control and Communications (SIBCON), Moscow, Russia, 2016.

[20] Y. Amit, D. Geman, K. Wilder, Joint induction of shape features and tree classifiers, IEEE Trans. Pattern Anal. Mach. Intell. 19 (11) (1997).

[21] J.R. Bergen, P. Anandan, K. Hanna, R. Hingorani, Hierarchical model-based motion esitimation, in: Proc. Eur. Conference Computer vison, 1002, European Conference on Computer Vision ECCV 1992: Computer Vision — ECCV'92, pp 237-252, May 2005.

[22] B. Bollobás, Modern graph theory, Graduate Texts in Mathematics, vol. 184, Springer-Verlag, 1998.

[23] J.-Y. Bouguet, Pyramidal Implementation of the Lucas Kanade Feature Tracker, Description of the Algorithm, Intel Corporation, Microprocessro Researc Lab, 2000.
[24] W.B. Dunbar, R.M. Murray, Model predictive control of coordinate multi-vehicle formations, in: Proc. IEEE Conference on Decision and Control, Las Vegas, NV, December 2002.
[25] E. Frew, T. McGee, Z. Kim, X. Xiao, S. Jackson, M. Morimoto, et al., Vision-based road following using a small autonomous aircraft, in: Proc. IEEE Aerospace Conference, Big Sky, MT, March 2004.
[26] R. Kumar, H. Sawhney, S. Samarasekera, S. Hsu, H. Tao, Y. Guo, et al., Aerial video surveillance and exploitation, IEEE Proc. 89 (10) (2001).
[27] J. Lee, R. Huang, A. Vaughn, X. Xiao, K. Hedrick, M. Zennaro, et al., Strategies of path-planning for a UAV to track a ground vehicle, Proceeding in AINS Conference, Research Gate, January 2003.
[28] K. Lucas, An iterative image registration technique with an application to stereo vision, in: Proc. of 7th International Joint Conference on Artificial Intelligence (IJCAI), pp. 674–679, April 1981.
[29] J.L. Mundy, A. Heller, The evolution and testing of a model-based object recognition system, in: Proc. IEEE Int. Conf. Computer Vision, https://doi.org/10.1109/ICCV.1990.139530, ISBN: 0-8186-2057-9, August 2002.
[30] R. Olfati-Saber, R.M. Murray, Distributed cooperative control of multiple vehicle formations using structural potential functions, in: Proc. of the IFAC World Congress, IFAC Proceedings Volumes 35(1), https://doi.org/10.3182/20020721-6-ES-1901.00244, July 2002.

CHAPTER

# 7 TRENDS OF SOUND EVENT RECOGNITION IN AUDIO SURVEILLANCE: A RECENT REVIEW AND STUDY

**N. Shreyas, M. Venkatraman, S. Malini and S. Chandrakala**

*Department of Computer Science and Engineering, School of Computing, SASTRA Deemed to be University, Thanjavur, India*

## CHAPTER OUTLINE



## 7.1 INTRODUCTION

Sound is assumed to be a significant part of our environment. Sound recognition is an upfront subject in the present sound event recognition (SER) hypotheses and covers a rich assortment of applications. This includes finding and grouping sound from genuine conditions like infant crying, individuals strolling, and dog yapping. Critical applications include detecting dangerous events/crimes [1]. It also helps in context understanding for robots [2] and cars [3]. The objective of SER is to assess the begin time and end time of every occasion and give a printed descriptor to every occasion inside a sound chronicle. Sounds can be classified into stationary and nonstationary [4].

The Cognitive Approach in Cloud Computing and Internet of Things Technologies for Surveillance Tracking Systems.
DOI: https://doi.org/10.1016/B978-0-12-816385-6.00007-6

Environmental sounds incorporate into the segment of nonstationary partition based on the fact that the state of sounds changes all of a sudden inside a brief period. In general, the conventional discourse acknowledgment methodologies may not be suitable for the different classes of environmental sounds. Numerous studies focus on finding a better and accurate extraction technique to extract more discriminative features. The basic classification is deciding whether the sound falls under the category of being either an indoor sound or an outdoor sound. Such examples of indoor classification include a living room, a bedroom, a classroom, an office, a childcare center, and so on. On looking at the outdoor classification, it includes a noisy street, a quiet residential neighborhood, bazaar full of people, and a school.

In the previous decade, diverse methodologies were proposed for SER task [5–13]. For instance, a few SER frameworks have addressed polyphonic identification utilizing the Gaussian blend models with shrouded Markov models (GMM-HMM) [14,15] and a couple of nonnegative lattice factorization. Various profound learning approaches have been proposed for SER and considered the front-line technique for SER.

Some of the sound recognition methods have produced groundbreaking results, for example, automatic speech recognition (ASR) [16,17] and music information retrieval (MIR) [18,19]. Environmental sound classification (ESC) is another essential branch of sound recognition and is broadly connected in reconnaissance, home computerization, scene examination, and machine hearing. Music and sound events differ much with an extensive variety of frequencies and are hardly all-around characterized, thus making ESC assignments more troublesome than MIR and ASR. Thus ESC still faces basic plan issues in performance and accuracy improvement. There exists a large acoustic variation present in each class. There are also problems of overlapping in the case of audio surveillance. In addition, reverberation poses a great challenge. Overview of SER is shown in Fig. 7.1.

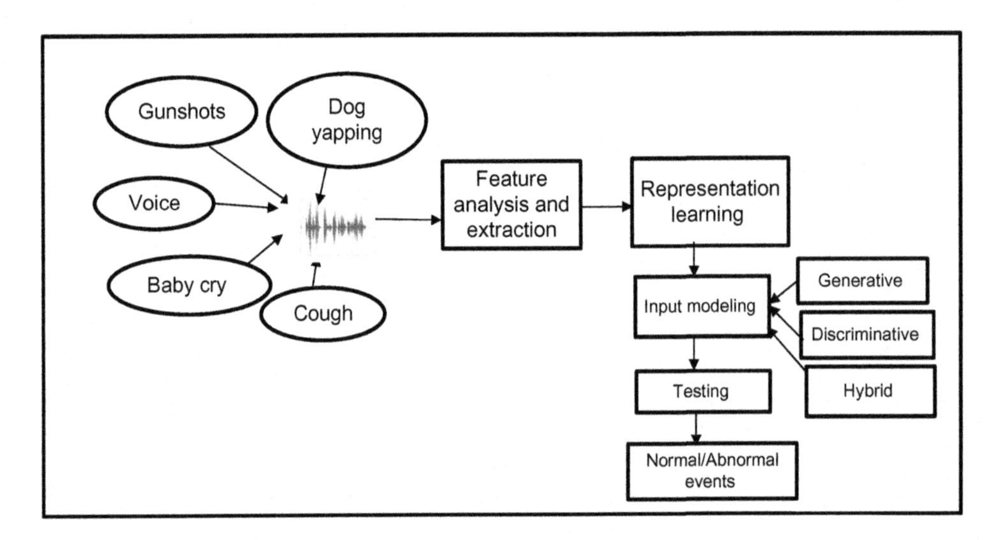


**FIGURE 7.1**

Overview of sound event recognition.

In this paper, we tend to perform a study on various methods by experimenting with three standard classification techniques. This is done by evaluating the methods over the publicly available benchmark data sets. The methods include support vector machines (SVMs) under nonprobabilistic linear classifier. Under deep learning methodologies, experimentation is done with neural networks (NNs) and convolutional neural network (CNN). Performance of each method is discussed in Section V.

## 7.2 NATURE OF SOUND EVENT DATA

Characteristics and various sound events to be recognized are elaborated with the help of the ESC-50 data set.

### 7.2.1 NATURE OF DATA

The data set used for this study is the ESC-50. This data set is available in HARVARD Data verse by the Warsaw University of Technology. It is a collection of about 2000 short environmental recordings available in a uniform format, which are suitable for the benchmarking methods of environment and sound classification. The clips are divided into 50 classes, where each class contains 40 clips. The data set contains stereo recordings from five major categories:

1. Natural soundscapes and water sounds, which includes rain, wind, water drops, thunderstorm, pouring water, sea waves, etc.
2. Human and nonspeech sounds like Snoring, laughing, footsteps, clapping, breathing, drinking, etc.
3. Interior or domestic sounds like door knock, vacuum cleaner, glass breaking, mouse click, keyboard typing, alarm clock, etc.
4. Exterior or urban noises like engine, church bells, train, siren, car horn, airplane, etc.
5. Animal sounds like dog, sheep, hen, frog, cat, cow, etc.

Each recording is 5-s long. These recordings were recorded in different streets and homes. The ESC-50 data set has been prearranged into fivefolds to perform easily a comparable cross-validation, making sure that the fragments from the same original source file are contained in a single grid. The data set is available under the terms of the Creative Commons Attribution Noncommercial license.

## 7.3 FEATURE EXTRACTION TECHNIQUES

The most important aspects in machine learning such as “feature selection” and “feature extraction” are discussed in the following section.

### 7.3.1 FEATURE SELECTION

It is very critical to understand and discriminate between the relevant and irrelevant parts of inputs. A problem of reducing the number of features used for modeling is referred to as dimensionality reduction. Especially, when dealing with a large number of variables, there is a need for dimensionality reduction. Feature selection can significantly improve a learning algorithms' performance

Mathematically speaking, given a set of features F = {f1, fi,. . ., fn}, the feature selection problem is to find a subset that maximizes the learner's ability to classify patterns. Formally, F' should maximize some scoring function. This general definition subsumes feature selection (i.e., a feature selection algorithm also performs a mapping but can only map to subsets of the input variables). The required number of samples (to achieve the same accuracy) grows exponentially with the number of variables. The classifier's performance usually will degrade for a large number of features. In many cases, the information that is lost by discarding variables is made up for by a more accurate mapping/sampling in the lower dimensional space. In practical cases, the number of training examples is fixed.

In theory, the goal is to find an optimal feature subset (one that maximizes the scoring function). In real-world applications, this is usually not possible. For most problems, it is computationally intractable to search the whole space of possible feature subsets. One usually has to settle for the approximations of the optimal subset. Most of the research in this area is devoted to finding efficient search-heuristics.

### 7.3.2 FEATURE EXTRACTION

There are various features that can be used while incorporating in SER. All these features take up the specific characteristics of the energy that is involved with the sound with the exception of the time-based features. Apart from time-based features, these energy-based features give us suitable predictable values. A few of the commonly used features are discussed below.

1. Time-domain based features

   Time-domain features are directly derived from the time-domain representation of a signal. There are various time-domain features, namely, waveform minimum and maximum, short-time energy, and zero-crossing rate.
2. Spectral-domain-based features

   Spectral-domain features are directly extracted from the power value of a spectrum. Some of the common spectral-domain features are fundamental frequency, pitch ratio, spectral moments, spectral flatness, spectral rolloff, spectral centroid, and bandwidth.
3. Cepstral-domain-based features

Cepstral-based features are features extracted with a cepstrum. The cepstrum is nothing but a nonlinear transformation of a spectrum. Most commonly used cepstrum-based features are Mel-frequency cepstral coefficients (MFCCs) and linear predictive cepstral coefficients.

a. Mel-frequency cepstral coefficients

   In sound processing, the Mel-frequency cepstrum [20] is a portrayal of the short-term power spectrum of a sound, based on a linear cosine transform of a log range spectrum on a nonlinear

Mel scale of frequency. The extraction usually involves applying the Fourier transform of the signal. Then, map the powers of the spectrum obtained above onto the Mel scale, using triangular overlapping windows. Finally, take the logs of the powers at each of the Mel frequencies followed by discrete cosine transforms. The resulting amplitude is MFCC representation of the sound signal.

**b.** Linear predictive cepstral coefficients

Linear predictive cepstral coefficients are the features that are extracted from the linear predictive coefficients by using a recursive technique. Linear predictive coefficients are coefficients that are obtained by minimizing the expectation of residual energy. It is obtained by two methods namely covariance and autocorrelation method. These features are computationally inexpensive to extract.

4. Energy-domain-based features

   Energy domain features extract energy from all the above three domains. It has two types of features, namely, log-energy first- and second-order derivatives and signal energy features.

5. Time-frequency-based features

   These features are extracted from a bidimensional function that connects the time and frequency. It includes the following features, namely, Gabor filter bank features, matching pursuit features, wavelet coefficient, spectrogram, wavelet coefficients, spectrogram image features, spectral variation, spectral flux, and histogram of oriented gradients features.

6. Perceptually driven-based features

These features are mainly intended for environmental sounds and events, because it is used to represent a nonstationary one. It includes features like Mel-frequency coefficients, log frequency coefficients, narrowband autocorrelation features, intonation, and Teager energy operator-based features, relative spectral perceptual linear prediction, perceptual linear prediction coefficients and derivatives, linear prediction coefficients and derivatives, gammatone cepstral coefficients, and spectral features based on gammatone filter bank.

## 7.4 SOUND EVENT RECOGNITION TECHNIQUES

Sound event data can be represented as a sequence of feature vectors and modeled using any sequence modeling techniques such as hidden-Markov model (HMM). The other way is to derive a global vector representation from a sequence of feature vectors and then modeled using any vector modeling techniques such as SVMs.

### 7.4.1 NONPROBABILISTIC LINEAR CLASSIFIER

#### *7.4.1.1 Support vector machines*

SVMs [21] is one of the nonprobabilistic linear classifiers that is supervised in nature with related learning calculations that break down information utilized for characterization and relapse investigation. SVMs are basic: The calculation makes a line, which isolates the classes in the event. The objective of the line is to expanding the edge between the focuses on either side of the purported choice line. One of the methods to choose the best hyperplane is selecting the one that represents the largest separation, or margin, between the two classes. We choose the hyperplane such that the distance from it to the nearest data point on each side is maximized. The advantage of this procedure is that after the detachment, the model can, without much of a stretch, figure the objective classes (names) for new cases. The relation governing the mapping of points x in the feature space into the hyperplane is given by the regularization parameter tells the SVM advancement. For substantial estimations of hyperparameter, the streamlining will pick a smaller edge hyperplane if that hyperplane completes a superior employment of getting all the preparation focuses ordered accurately. On the other hand, a little estimation of hyperparameter will cause the enhancer to search for a bigger edge isolating hyperplane, regardless of whether that hyperplane misclassifies or not. SVM is based on the concept of decision planes that define decision boundaries. A decision plane is one that separates between a set of objects having different class memberships. SVM views the input data as two sets of vectors in an $n$-dimensional space. The vectors that constrain the width of the margin are the support vectors. It constructs a separating hyperplane in that space, one of which maximizes the margin between the two data sets. To calculate the margin, two parallel hyperplanes are constructed, one on each side of the separating hyperplane, which is pushed up against the two data sets. A good separation is achieved by the hyperplane that has the largest distance to the neighboring data points of both classes, since, in general, the larger the margin, the lower the generalization error of the classifier.

#### *7.4.1.2 Hidden-Markov model*

HMM is one of the generative model-based approaches used to model the temporal variations in the environmental audio scenes with noise. It takes the sequence of feature vectors as input and produces a hard target output. It mainly has three parameters, namely, transition probability, states, and mixtures. This model for a class is trained to maximize the likelihood of a model producing the sequence of input of that class. A test scene is fed as input in order to get the probability for that sequence provided by the model, find the maximum one, and assign it to that test event. Various parameters like the number of mixtures and states influence the recognition accuracy.

### 7.4.2 DEEP LEARNING METHODOLOGIES

Deep learning is a set of algorithms, which is part of a broader family of machine learning methods, which is based on the representations of data learning. The structure, functions, and working of information processing and communication patterns present in the biological nervous systems of the brain have been an inspiration, which gave rise to the development of deep learning. Deep learning has undergone a dramatic improvement in addressing important and complex problems like computer vision, natural language processing, social network filtering, machine translation, medical image analysis, material inspection, and sound recognition majorly in the form of speech. It can design powerful abstractions in the input through the use of a variety of architectures that are composed of more than one nonlinear transformations. An audio signal can be characterized as a vector and, therefore, can be handled by multiple standard methods. Sound event data can be represented as a sequence of feature vectors and modeled using any deep learning-based sequence modeling techniques such as a recurrent neural network (RNN). The other way is to derive a global vector representation from a sequence of feature vectors and then modeled using any deep learning-based vector modeling techniques such as deep neural network (DNN) and CNN. This section explains deep learning methodologies that include methods like NNs, CNNs, and RNNs.

#### *7.4.2.1 Neural networks*

NNs, otherwise called artificial neural systems, are propelled by impersonating the manner in which the human mind works. This network is in the form of a circuit or network, which is composed of a group or groups of artificial neurons/nodes that use a mathematical as well as a computational model for processing of information, which is based on a connectionist. The principal structure of an NN is sorted out in layers. Layers are constituted of various interconnected neurons or nodes, which contain initiation capacities. The connections between the neurons are modeled as weights. Every neuron essentially comprises data sources that are duplicated by connection weights and after that registered by a numerical capacity, which decides the enactment of the neuron. All inputs are modified by performing a linear combination using weights. Finally, the amplitude of the output is controlled by an activation function.

In spite of the fact that NNs have been embraced in numerous applications, it demonstrates a few deficiencies while applying for spatial and transient structure information for pictures, sound, discourse, and content. Right off the bat, the outline of NNs is completely connected layers that have a substantial number of parameters and the number of parameters quickly increments amid the preparing process. It prompts moderate learning for spatial and worldly structure information. In addition, every combination of neurons between two layers of NNs has their own parameters keeping the system abuses and highlights the connections in high-dimensional spatial and temporal context. Cakir et al. [22] suggested a method using three types of features, namely, MFCCs, Mel-band energies, and log Mel-band energies and DNNs with two hidden layers is used as the classifier. The model was tested with recordings from realistic everyday environments leading to an overall accuracy of 63.8%. The model has an accuracy higher than HMM-GMM [15] with a large margin of over 20%.

#### *7.4.2.2 Convolutional neural networks*

The CNNs are a sort of neural system design and were produced to beat the downsides of NNs when managing the spatial structure information. The CNN is most commonly used for analyzing visual imagery. The CNN was enlivened in light of preparing the visual cortex of people. A CNN comprises three essential segments—convolutional layers, pooling layers, and completely associated layers. In machine learning, a CNN can be treated as a type of feed-forward artificial neural network (ANN). A CNN is composed of a couple of distinct layers of minor neuron collections, which are stacked onto each other. Commonly used distinct types of layers are convolutional, pooling, Rectified linear unit (ReLU), fully connected layer, and loss layer. Convolutional networks use relatively little preprocessing when compared with other classification algorithms, because in this case, the network learns the filters rather than hard-engineering. The above-mentioned collections process portions of the input data as a consequence that the outputs of these collections are then tiled so that they come one over the other to obtain a neater depiction of the original input. This is often echoed for every similar layer. Just like any other NN, we use an activation function to make our output nonlinear. In the case of a CNN, the output of the convolution will be passed through the ReLU function.

ReLU: $f(x) = \max(0, x)$

This function performs the effective removal of negative values from the activation map by setting all these values to zero. The applications of CNN are majorly in image and video recognition, recommender systems, image and audio classification, medical image analysis, and natural language processing.

A variation in the CNN is the deep Q-network, which is a type of deep learning model that is a combination of a deep CNN with Q-learning. The advantage is this method can learn directly from high-dimensional sensory inputs. Another variation is the convolutional deep belief networks. The core difference is that they provide a generic structure that can be used in many image and signal processing tasks. Benchmark results on standard data sets have been obtained using this method.

A method was suggested by Piczak [19] by squaring the frequency area under the log Mel spectrogram. This increased the accuracy by 20.5% on the ESC 50 data set. Log Mel spectrogram was again used by Takahashi et al. [23]. Agarwal et al. [24] implemented the gamma tone spectrogram, used in the CNN similar to Piczak [19] and said to have achieved an accuracy 79.1 % in the ESC 50 data set. Dai et al. [25] suggested a deep CNN model with 1-dimensional (1-D) Convolutional layers using 1-D raw data as input and it proved the purpose by attaining good accuracy with CNN using log Mel spectrogram inputs [26].

#### *7.4.2.3 Recurrent neural network*

RNN is one of the deep learning models that are used for modeling the arbitrary length sequences by applying a transition function to all its hidden states in a recursive manner. It is well suited for sequence modeling techniques related to the time variations as well as the time-invariant inputs. It has a fixed structure architecture, but it is differentiable from all the nodes [27]. The derivative of the loss function is calculated for each and every parameter. It has many cycles in its structure, so it is well suited for the time-varying inputs. These cycles are used to model the temporal variations present in the input. The activation function followed by this RNNs hidden state is a function that depends on its previous states only [28]. It maps the sequence of inputs into the fixed size vector, and then, it is fed as an input to a softmax activation function and it produces the output. The RNN overcomes the long-term dependences, because it depends on the current input and the previous output. The problem over RNN is exploding and vanishing gradients due to its more number of transitions.

## 7.5 EXPERIMENTATION AND PERFORMANCE ANALYSIS

In audio surveillance, SER is an important task. In this study, we review various methods and features used for that task. In this section, we have done experimentation over the ESC-50 data set using MFCC features with CNN, ANN, and SVM as classifiers and the results are tabulated in Table 7.1. The best performance is acknowledged for deep learning-based approaches. MFCC-based deep learning approaches of our work are compared with various literature works of different features like Logmel [19] and Gammatone [24].

**Table 7.1 Comparative Study on ESC-50 Data set**

| S. No. | Methods Used | Accuracy(%) |
|---|---|---|
| 1 | **MFCC + SVM** | **59.5** |
| 2 | **MFCC + ANN** | **74.84** |
| 3 | **MFCC + CNN** | **78.54** |
| 4 | Logmel – CNN [19] | 64.5 |
| 5 | MFCC – DNN [15] | 63.8 |
| 6 | Gammatone–CNN [24] | 79.1 |

*ANN*, Artificial neural network; *CNN*, convolutional neural network; *DNN*, deep neural network; *MFCC*, Mel-frequency cepstral coefficient; *SVM*, support vector machine.

### 7.5.1 DATA SET

There are various data sets for SER tasks, which are publicly available for experimentation. Few of them are Challenge-DCASE2013, TUT-DCASE2016, ESC-10, UrbanSound8k, CICESE, and ESC-50. We use ESC-50 for our experimentation.

### 7.5.2 COMPARATIVE STUDY ON RELATED WORK

Various methods evaluated over ESC-50 data set are studied, and comparison results are shown in Table 7.1. Piczak [19] proposed a model using SVM. In this model, two types of features were extracted: zero-crossing rate and MFCCs. This shows an overall accuracy of 39.6%. The proposed model in this paper has an accuracy of 59.5 %. The model outperforms the former by 19.9%. Heittola et al. [15] proposed a multilabel feedforward DNNs for the SER task. The model is evaluated with recordings from realistic everyday environments, and the achieved overall accuracy is 63.8%. The model studied in this paper has an accuracy of 74.84% and outperforms HMM-GMM with a margin of 11.04%. Jeong et al. [6] show a good performance when applied the CNN for the SER task. The work uses both short-term and long-term audio signals as input features to feed into convolutional network architectures. They used the 1-D convolution layer with 64 filters. The proposed model obtained better results than the baseline system in terms of score and accuracy. The model has achieved an accuracy of 78.54%.

Our methods have shown reasonable recognition accuracy over other methods mentioned in Table 7.1. From our experimental study, we observe that the performance of the ANN is better than the other methods using SVM as classifier. The CNN outperforms both SVM and ANN methods. Furthermore, on the analysis with the help of the confusion matrix, we observe that sounds that are confused over human perception is also misclassified by the system.

Further, on studying the manipulations over the number of layers of the neural networks, we find that the accuracy of the model goes down. This is natural as we do not know the trends in neural models as the number of layers and the number of nodes in each layer go up. Sometimes, the accuracy may increase or decrease according to the nature of the sound classes we take upon. For example, in the ANNs on increasing the number of layers by 2, the accuracy drops by more than 10%.

## 7.6 FUTURE DIRECTIONS AND CONCLUSION

More research is needed on understanding the characteristics of the SVM, NN, and CNN techniques applied over the sound classification task under highly clamorous conditions. For successfully developing of such a sound recognition system, it is essential to take into account such noisy disturbances. In order to achieve this, different kinds of techniques aiming essentially at finding robust and invariant signal features can be used with the help of adaptation methods or robust decision strategies. From the observations made so far, we suggest that additional efforts has to be taken to combine existing features like MFCC and LPCC to achieve improved results. Further research can be done over different data sets collected from some real-world scenarios to improve robustness. Many new model-driven approaches such as discriminative-model based approaches and hybrid-model based approaches can be applied to achieve better results. Our future studies will be carried over long duration tasks, since most of the above-mentioned methods are suitable only for short-duration tasks.

SER is one of the most important tasks in audio surveillance. In this survey, we focus on traditional and deep learning approaches for SER task over the ESC-50 data set. We have done a comparative study of traditional and deep learning approaches with some methods available from the literature. With this analyzed results of our study, we conclude that deep learning approaches work well for the SER tasks.

## REFERENCES

[1] P. Foggia, N. Petkov, A. Saggese, N. Strisciuglio, M. Vento, Audio surveillance of roads: a system for detecting anomalous sounds, IEEE Transactions on Intelligent Transportation Systems 17 (1) (2015) 279–288.

[2] H.M. Do, W. Sheng, M. Liu, S. Zhang, Context-aware sound event recognition for home service robots, 2016 IEEE International Conference on Automation Science and Engineering (CASE), IEEE, 2016, pp. 739–744.

[3] S. Singh, S.R. Payne, P.A. Jennings, Toward a methodology for assessing electric vehicle exterior sounds, IEEE Transactions on Intelligent Transportation Systems 15 (4) (2014) 1790–1800.

[4] J.S. Hu, W.H. Liu, Location classification of nonstationary sound sources using binaural room distribution patterns, IEEE Transactions on Audio, Speech, and Language Processing 17 (4) (2009) 682–692.

[5] E. Cakir, T. Heittola, H. Huttunen, T. Virtanen, Polyphonic sound event detection using multi label deep neural networks, 2015 International Joint Conference on Neural Networks (IJCNN), IEEE, 2015, pp. 1–7.

[6] I.Y. Jeong, S. Lee, Y. Han, K. Lee, Audio event detection using multiple-input convolutional neural network. Detection and Classification of Acoustic Scenes and Events (DCASE) (2017).

[7] Y. Chen, Y. Zhang, Z. Duan, DCASE2017 sound event detection using convolutional neural network, Detection and Classification of Acoustic Scenes and Events (2017).

[8] S. Adavanne, G. Parascandolo, P. Pertilä, T. Heittola, T. Virtanen, Sound event detection in multichannel audio using spatial and harmonic features. arXiv preprint arXiv:1706.02293 (2017).

[9] T.H. Vu, J.C. Wang, Acoustic scene and event recognition using recurrent neural networks, Detection and Classification of Acoustic Scenes and Events 2016 (2016).

[10] G. Parascandolo, H. Huttunen, T. Virtanen, Recurrent neural networks for polyphonic sound event detection in real life recordings, in: 2016 IEEE International Conference on Acoustics, Speech and Signal Processing (ICASSP), 2016, pp. 6440–6444.
[11] R. Lu, Z. Duan, Bidirectional GRU for sound event detection, Detection and Classification of Acoustic Scenes and Events (2017).
[12] J. Zhou, Sound event detection in multichannel audio LSTM network, Proc. Detection Classification Acoust. Scenes Events, 2017.
[13] M. Zöhrer, F. Pernkopf, Gated recurrent networks applied to acoustic scene classification and acoustic event detection, Detection and Classification of Acoustic Scenes and Events 2016 (2016).
[14] A. Mesaros, T. Heittola, A. Eronen, T. Virtanen, Acoustic event detection in real-life recordings, in: Proceedings of the 18th European Signal Processing Conference, Aalborg, Denmark, 2010, pp. 1267–1271.
[15] T. Heittola, A. Mesaros, A.J. Eronen, T. Virtanen, Context-dependent sound event detection, EURASIP J. Audio Speech Music Process. 1 (2013) 1–13.
[16] A. Graves, A.R. Mohamed, G. Hinton, Speech recognition with deep recurrent neural networks, in: Acoustics, Speech and Signal Processing (ICASSP), 2013 IEEE International Conference, IEEE, 2013, pp. 6645–6649.
[17] G. Hinton, L. Deng, D. Yu, G.E. Dahl, A.R. Mohamed, N. Jaitly, et al., Deep neural networks for acoustic modeling in speech recognition: the shared views of four research groups, IEEE Signal. Process. Mag. 29 (6) (2012) 82–97.
[18] M.A. Casey, R. Veltkamp, M. Goto, M. Leman, C. Rhodes, M. Slaney, Content based music information retrieval: current directions and future challenges, Proc. IEEE 96 (4) (2008) 668–696.
[19] K.J. Piczak, Esc: dataset for environmental sound classification, in: ACM International Conference on Multimedia, 2015, pp. 1015–1018.
[20] S. Ntalampiras, I. Potamitis, N. Fakotakis, Probabilistic novelty detection for acoustic surveillance under real-world conditions, IEEE Transactions on Multimedia 13 (4) (2011) 713–719.
[21] G. Georgoulas, V.C. Georgopoulos, C.D. Stylios, Speech sound classification and detection of articulation disorders with support vector machines and wavelets, 2006 International Conference of the IEEE Engineering in Medicine and Biology Society, IEEE, 2006, pp. 2199–2202.
[22] E. Cakir, T. Heittola, H. Huttunen, T. Virtanen, Polyphonic sound event detection using multi label deep neural networks, 2015 International Joint Conference on Neural Networks (IJCNN), IEEE, 2015, pp. 1–7.
[23] N. Takahashi, M. Gygli, B. Pfister, L. Van Gool, Deep convolutional neural networks and data augmentation for acoustic event detection. arXiv preprint arXiv:1604.07160.
[24] D.M. Agarwal, H.B. Sailor, M.H. Soni, H.A. Patil, Novel teo based gammon features for environmental sound classification, in: 2017 25th European Signal Processing Conference (EUSIPCO), IEEE, 2017, pp. 1809–1813.
[25] W. Dai, C. Dai, S. Qu, J. Li, S. Das, Very deep convolutional neural networks for raw waveforms, in: Acoustics, Speech and Signal Processing (ICASSP), 2017 IEEE International Conference, IEEE, 2017, pp. 421–425.
[26] D.M. Agrawal, H.B. Sailor, M.H. Soni, H.A. Patil, Novel teo-based gammatone features for environmental sound classification, in: 2017 25th European Signal Processing Conference (EUSIPCO), IEEE, 2017, pp. 1809–1813.
[27] Z.C. Lipton, J. Berkowitz, C. Elkan, A critical review of recurrent neural networks for sequence learning. In: arXiv: 1506.00019v4 [cs. LG], 2015, pp. 1–33.
[28] P. Liu, X. Qiu, X. Huang, Recurrent neural network for text classification with multi-task learning, in: 25th International Joint Conference on Artificial Intelligence (IJCAI), 2016, pp. 2873–2879.

## FURTHER READING

S. Mun, S. Shon, W. Kim, D.K. Han, H. Ko, Deep neural network based learning and transferring mid-level audio features for acoustic scene classification, in: Acoustics, Speech and Signal Processing (ICASSP), 2017 IEEE International Conference, IEEE, 2017, pp. 796–800.

A. Fadeev, O. Missaoui, H. Frigui, Dominant audio descriptors for audio classification and retrieval, in: International Conference on Machine Learning and Applications, 2009.

CHAPTER

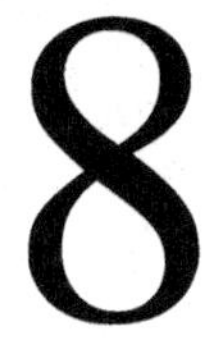

# OBJECT CLASSIFICATION OF REMOTE SENSING IMAGE USING DEEP CONVOLUTIONAL NEURAL NETWORK

**P. Deepan and L.R. Sudha**

*Department of Computer Science and Engineering, Annamalai University, Chidambaram, India*

**CHAPTER OUTLINE**



## 8.1 INTRODUCTION

Image classification plays an important role in remote sensing images and is used for various applications such as environmental change, agriculture, land use/land planning, urban planning, surveillance, geographic mapping, disaster control, and object detection and also it has become a hot research topic in the remote sensing community [1]. The remote sensing image data can be obtained from various resources like satellites, airplanes, and aerial vehicles. Earlier, the spatial satellite image resolution was used, which was very low, and the pixel sizes were typically coarser and the image analysis methods for remote sensing images are based on pixel-based analysis or subpixel analysis for this conversion [2]. In this case, sometimes it is difficult to classify the scene images at pixel level clearly.

In order to solve this problem, some researchers have focused on object-based image analysis instead of individual pixels [3]. Here, the term "objects" represents meaningful scene components that distinguish an image. The object-level methods gave better results of image analysis than the pixel-level methods. With the development of machine learning algorithm, the semantic-level

The Cognitive Approach in Cloud Computing and Internet of Things Technologies for Surveillance Tracking Systems.
DOI: https://doi.org/10.1016/B978-0-12-816385-6.00008-8

method is also used for analyzing the remote sensing image [4]. The semantic-level image classification aims to provide the label for each scene image with a specific semantic class. The scene images are manually extracted from the large-scale remote sensing image, for example, airplane, beach, forest, road, and river [3,4].

A wide number of techniques have been developed for object classification [1]. In general, the object classification methods are divided into three categories based on the features they use, namely, handcraft feature learning method, unsupervised feature learning method, and deep feature learning-based method [5]. Earlier, scene classification was based on the handcraft feature learning-based method. This method [6,7] was mainly used for designing the engineering features, such as color, shape, texture, and spatial and spectral information. The unsupervised feature learning method [8] is an alternative for the handcrafted feature method and training the unlabeled data for remote sensing image classification. The aim of the unsupervised feature learning method is used to identify the low-dimensional features that capture some underlying high-dimensional input data. When the feature learning is performed in an unsupervised way, it enables a form of semisupervised learning, where features learned from an unlabeled data set are then employed to improve performance in a supervised setting with labeled data. There are several unsupervised feature learning methods available such as k-means clustering, principal component analysis (PCA), sparse coding, and autoencoding. In real-time applications, the unsupervised feature learning methods have achieved high performance for classification compared with handcrafted-feature learning methods [9]. However, the lack of semantic information provided by the category label cannot promise the best discrimination between the classes. So we need to improve the classification performance and to extract powerful discriminant features for improving classification performance.

Deep learning [10] is a powerful machine learning technique for solving a wide range of computer applications. It is composed of multiple processing layers that can learn more powerful feature representations of data with multiple levels of abstraction [11]. In the deep learning technique, a several number of models are available such as convolutional neural network (CNN), deep autoencoders, deep belief network (DBN), recurrent neural network (RNN), and long short-term memory (LSTM). The models are aimed to get high-level features. When compared with traditional methods, deep learning methods do not need manual annotation and knowledge experts for feature extraction. Feature extraction and classifications are combined together in this model.

This research paper has been organized as follows. Section 8.2 describes the review and related works for the scene classification. Section 8.3 discusses the visual geometry group (VGG)-16 deep CNN for scene classification. Section 8.4 provides detail description about the benchmark data set. Section 8.5 describes the experimental results and analysis. Finally, conclusions are shown in Section 8.6.

## 8.2 RELATED WORKS

Remote sensing images are more valuable tool for monitoring the Earth surface and mainly used for various applications such as surveillance, agricultural monitoring, metrology, mineralogy, and environmental science. A lot of classification methods have been proposed to deal with the remote sensing image classification. The traditional object classification is based on either supervised or unsupervised learning methods. Some researchers have proposed object classification methods using supervised learning techniques such as support vector machines (SVMs), artificial neural networks (ANNs), random forest, k-nearest neighbors, decision tree, and sparse representation classifier [1].

Bazi and Melgani [12] have introduced the genetic optimization framework in an SVM for classifying hyperspectral remote sensing images. Archibald and Fann [13] have proposed embedded feature-selection algorithm that is tailored to operate with SVMs to perform band selection and classification. The ANN is a nonparametric learning technique, which is a more sophisticated and robust method for image classification. The ANN produces higher accuracy from few training data. Kavzoglu and Mather [14] have proposed back-propagating ANNs in land cover classification. The Random forest is an ensemble classifier that produces multiple decision trees using randomly selected training samples. Belgiu and Drăguţ [15] have done a survey of object classification by using random forest techniques and their applications. Hayesa et al. [16] have introduced high-resolution land cover classification using Random forest.

Cheriyadat [17] proposed an unsupervised feature learning approach for scene classification. Unsupervised feature learning approach is to generate feature representation for various high-resolution aerial scenes. Chaib et al. [18] have introduced an informative feature selection method for high-resolution satellite image classification by using a PCA classifier. Zhang et al. [19] have developed a hybrid model satellite image classification using multifeature joint sparse coding. Sheng et al. [20] have proposed a sparse coding-based multifeature for image classification. But the sparse coding is more expensive when dealing with big data. Therefore the sparse feature coding can be used only for small-scale problems.

In recent years, a CNN has been used in various remote sensing applications, such as object classification, land use scene classification, and object detection. The CNNs is a one type of ANNs, which consist of a series of layer such as convolution, subsampling/pooling, fully connected, and softmax function. However, additional layers such as dropout, batch normalization, and optimization can be used for avoiding overfitting problem and improving the generalization of the model. The last layer depends on the problem types, for binary classification sigmoid is used. For multi-class classification, softmax function is used.

Zhong et al. [21] proposed a novel approach for scene classification of high spatial resolution imagery by using large patch CNNs. The large patch sampling is used to generate hundreds of possible scene patches for feature learning. Almost all the existing CNN models train only small data sets. In order to solve this problem, deep CNN is introduced, which can train large data sets. Kruithof et al. [22] have proposed an object recognition using deep CNNs with complete transfer and partial frozen layers. Zou et al. have developed a deep learning-based feature selection for remote sensing scene classification. The popular deep-learning technique, that is, the DBN, has achieved feature abstraction by minimizing the reconstruction error over the whole feature set, and features with smaller reconstruction errors would hold more features intrinsic for image representation [23]. Kussul et al. [24] have introduced deep learning classification of land cover and crop types using remote sensing data. While analyzing the literature, we found that the classification performance of all the above techniques is not satisfactory. So in order to improve the performance, we have included the VGG-16 deep CNN in our proposed approach.

## 8.3 VGG-16 DEEP CONVOLUTIONAL NEURAL NETWORK MODEL

This section describes the VGG-16 deep CNN model for classification of remote sensing images using deep CNNs. The model is one of most powerful deep CNN, which was proposed by Simonyan and Zisserman. Fig. 8.1 shows the architecture diagram of the VGG-16 deep CNN model. It takes input image at the low level and processes them through a sequence of computational units and obtains the necessary values for classification in the higher layer. This model consists of 13 convolutional layers with $3 \times 3$ filter size, five subsampling/max pooling layer with a size of $2 \times 2$, and two fully connected layers with activation function and softmax function.

The convolutional layers extract features from the input images. The 13 convolutional layers are distributed in five blocks. The first two blocks contain two convolutional layers in each block. Similarly, the remaining three blocks consist of three convolutional layers in each block. The first block convolutional layer extracts low-level features such as lines and edges. Higher level layer extracts high-level features. Every convolutional filter has a kernel size of $3 \times 3$ filters with stride 1. The filter size of the convolutional layer is gradually increased from 64 to 512.

The subsampling layer is used to reduce the feature resolution. This layer reduces the number of connection between the convolutional layers, so it will lower the computational time also. There are three types of pooling: max pooling, min pooling, and average pooling. In each case, the input image is divided into nonoverlapping two-dimensional spaces. For example, if the input size is

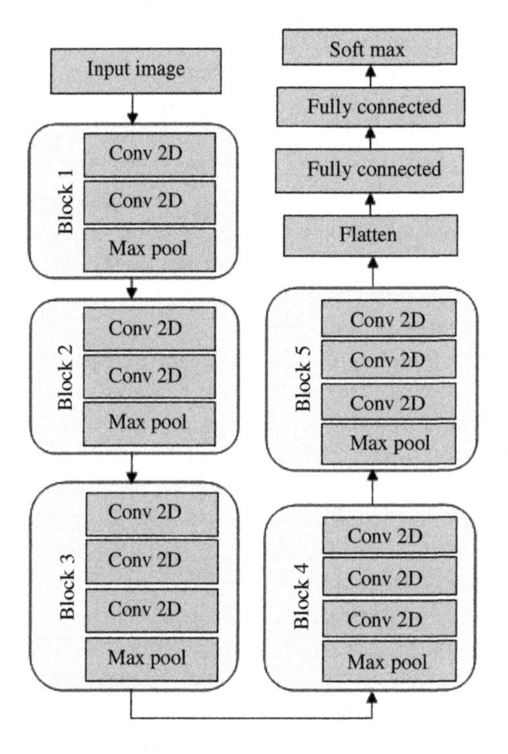


**FIGURE 8.1**

Architecture of visual geometry group-16 deep convolutional neural network model.

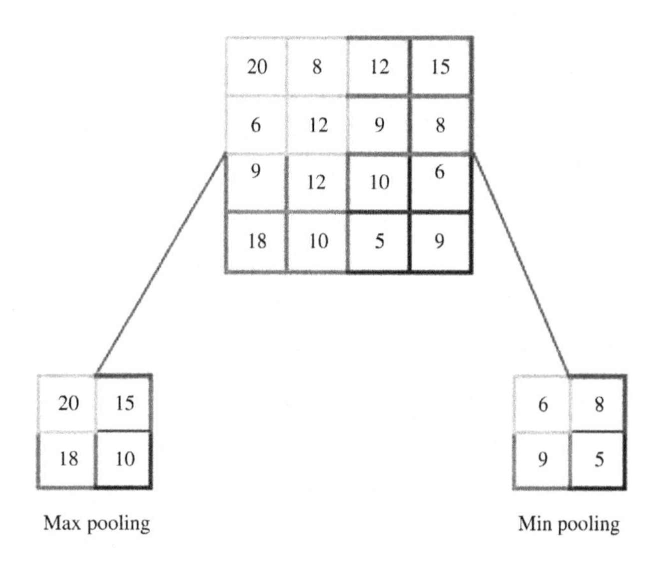


**FIGURE 8.2**

Pictorial representation of max pooling and average pooling processes.

$4 \times 4$ and the subsampling size is $2 \times 2$, a $4 \times 4$ image is divided into four nonoverlapping of matrices $2 \times 2$. For max pooling, the maximum value of the four values is selected. In the case of min pooling, the minimum value of the four values is selected. Fig. 8.2 shows the operation of the max pooling and average pooling processes.

The VGG-16 deep CNN model ends with two fully connected layer and softmax function. In these layers, sum of all the weights of previous layer features is calculated and the specific output is determined. Finally, fully connected layers reduce the dimension into 4096 and classify the 10 class object using softmax function. The activation function improves the deep CNN performance. In this chapter, three standard activation functions, such as tanh, eLu, and rectified linear unit (ReLU), have been used. The ReLU is one of the standard and popular activation functions in the last few years. The ReLU activation function is defined as:

$$b_{i,j,k} = \max\left(a_{i,j,k}, 0\right) \tag{8.1}$$

where $a_{i,j,k}$ is the input of the activation function at location $(i, j)$ on the $k$th channel. In this layer, we remove every negative value from the filtered images and replace it with zeros. Fig. 8.3 elaborates the process of activation function.

The overfitting is an unneglectable problem in the VGG-16 deep CNN model that can be reduced by regularization. In this chapter, we use the effective regularization technique Dropout. The Dropout was introduced by Hinton et al. [25] and it has been proved effective in reducing overfitting. The dropout techniques are used in the fully connected layer and we can specify the different level of dropout parameters like 0.2, 0.3, and 0.5. The VGG-16 deep CNN model was trained with backpropogation algorithm and root mean square property (RMSprop). The RMSprop is used to reduce the loss function of the VGG-16 deep CNN model. Table 8.1 summarizes the hyperparameters of the VGG-16 deep CNN model.

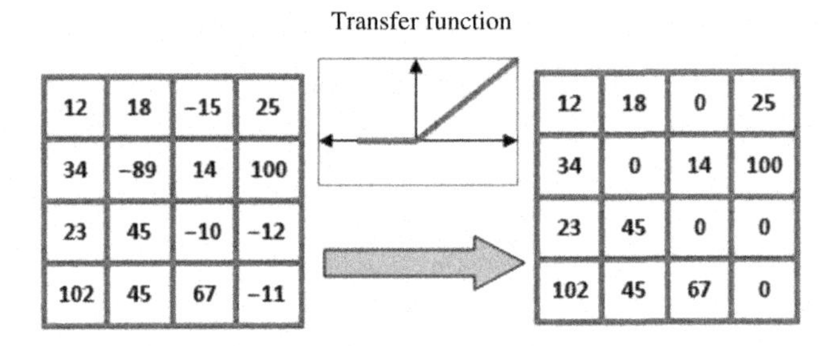


**FIGURE 8.3**

Pictorial representation of activation function.

**Table 8.1 Summary of VGG-16 Model**

| Layer (Type) | Output Shape | Parameter |
|---|---|---|
| input_1 (Input Layer) | (None, 224, 224, 3) | 0 |
| b1_conv1 (Conv2D) | (None, 224, 224, 64) | 1792 |
| b1_conv2 (Conv2D) | (None, 224, 224, 64) | 36,928 |
| b1_pool (MaxPooling2D) | (None, 112, 112, 64) | 0 |
| b2_conv1 (Conv2D) | (None, 112, 112, 128) | 73,856 |
| b2_conv2 (Conv2D) | (None, 112, 112, 128) | 14,7584 |
| b2_pool (MaxPooling2D) | (None, 56, 56, 128) | 0 |
| b3_conv1 (Conv2D) | (None, 56, 56, 256) | 295,168 |
| b3_conv2 (Conv2D) | (None, 56, 56, 256) | 590,080 |
| b3_conv3 (Conv2D) | (None, 56, 56, 256) | 590,080 |
| b3_pool (MaxPooling2D) | (None, 28, 28, 256) | 0 |
| b4_conv1 (Conv2D) | (None, 28, 28, 512) | 1,180,160 |
| b4_conv2 (Conv2D) | (None, 28, 28, 512) | 2,359,808 |
| b4_conv3 (Conv2D) | (None, 28, 28, 512) | 2,359,808 |
| b4_pool (MaxPooling2D) | (None, 14, 14, 512) | 0 |
| b5_conv1 (Conv2D) | (None, 14, 14, 512) | 2,359,808 |
| b5_conv2 (Conv2D) | (None, 14, 14, 512) | 2,359,808 |
| b5_conv3 (Conv2D) | (None, 14, 14, 512) | 2,359,808 |
| b5_pool (MaxPooling2D) | (None, 7, 7, 512) | 0 |
| flatten_1 (Flatten) | (None, 25,088) | 0 |
| dense_1 (Dense) | (None, 1024) | 25,691,136 |
| dropout_1 (Dropout) | (None, 1024) | 0 |
| dense_2 (Dense) | (None, 10) | 10,250 |

Total parameters: 40,416,074
Trainable parameters: 32,780,810
Nontrainable parameters: 7,635,264

VGG, *Visual geometry group.*

## 8.4 DATA SET DESCRIPTION

The North Western Polytechnical University (NWPU)-RESISC 45 class data set is the publicly available benchmark data set. It is mainly used for remote sensing image scene classification and created by the NWPU. The remote sensing data set is extracted from Google Earth. The NWPU-RESISC 45 class data set contains airplane, airport, baseball diamond, basketball court, beach, bridge, chaparral, church, circular farmland, cloud, commercial area, dense residential, desert, forest, freeway, golf course, ground track field, harbor, industrial area, intersection, island, lake, meadow, medium residential, mobile home park, mountain, overpass, palace, parking lot, railway, railway station, rectangular farmland, river, roundabout, runway, sea ice, ship, snow berg, sparse residential, stadium, storage tank, tennis court, terrace, thermal power station, and wetland. The data set contains 31,500 satellite images and each class consists of 700 images. The resolution of each image has size of $256 \times 256$ pixels with RGB color space. In our proposed method, we randomly select 10 classes such as airplane, beach, commercial area, desert, forest, lake, overpass, river, tennis court, and wetland.

Fig. 8.4 shows some example images from the NWPU-RESISC 45 class data set for image classification. In the VGG-16 deep CNN model, the data set has been split into training, validation, and testing data sets separately. The training and testing set descriptions are shown in Table 8.2. The validation images are randomly selected from training sample based on the size of validation.

## 8.5 EXPERIMENTAL RESULTS AND ANALYSIS

Experiments are conducted on 10 classes such as airplane, beach, commercial area, desert, forest, lake, overpass, river, tennis court, and wetland from the NWPU-RESISC 45 class data set using VGG-16 deep CNN. The VGG-16 model was trained and tested with 7000 images using tensor flow in Core i7 CPU 2.6 GHz, 1-TB hard disk, and 8-GB RAM.

In order to evaluate the performance of proposed work, six common evaluation metrics such as accuracy, precision, recall, F1-score, confusion matrix, and receiver operating characteristics (ROC) curve are calculated using the formula given in Eqs. (8.2)–(8.5):

$$\text{ACCURACY} = \frac{\text{Total No. of Correct Prediction}}{\text{No. of Input Samples}} \tag{8.2}$$

$$\text{PRE}_i = \frac{\text{TP}_i}{\text{TP}_i + \text{FP}_i} \tag{8.3}$$

$$\text{REC}_i = \frac{\text{TP}_i}{\text{TP}_i + \text{FN}_i} \tag{8.4}$$

$$\text{F}_1^i = 2 \times \frac{\text{PRE}_i \times \text{REC}_i}{\text{PRE}_i + \text{REC}_i} \tag{8.5}$$

where TP (true positives) denotes the number of correct detections of the object, TN (true negatives) denotes the number of wrong detections of the object, FN (false negatives) denotes the number of missed objects, and FP (false positives) denotes the number of incorrect detections of the

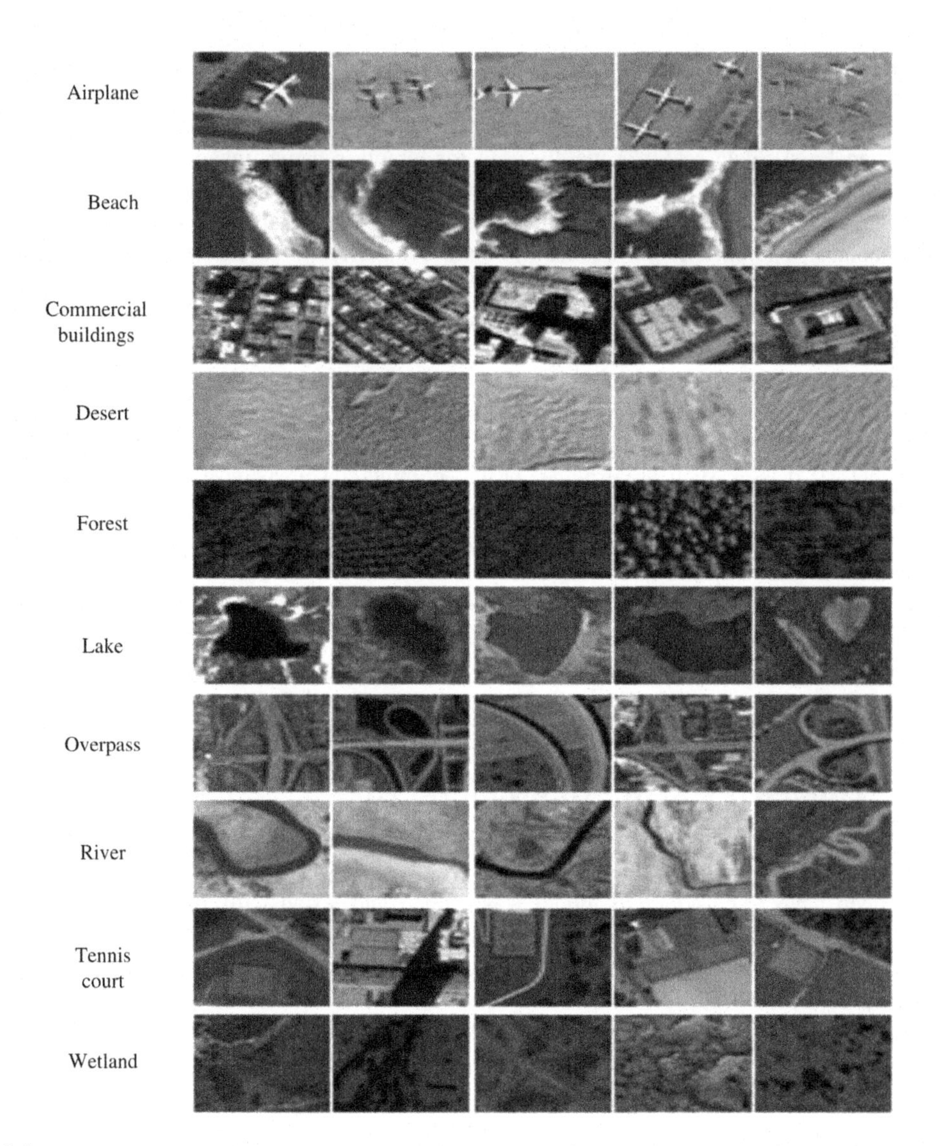


**FIGURE 8.4**

Image examples of different classes from the North Western Polytechnical University-RESISC 45 class data set.

object. The average accuracy, precision, recall, F1-score, and loss of VGG-16 model on validation data set were 0.947, 0.95, 0.95, and 0.2341, respectively. The results suggest that the VGG-16 model could classify all the classes with higher accuracy. The wetland class gives an average of 0.82 results due to shadow class and the class is missclassified as river, lake, and forest. The evaluation metrics of the proposed VGG-16 deep CNN model was shown in Table 8.3.

**Table 8.2 Statistics of Training and Validation Data Set**

| Class Number | Class Name | No. of Training Sample | No. of Validation Sample | No. of Testing Sample |
|---|---|---|---|---|
| 0 | Airplane | 538 | 100 | 19 |
| 1 | Beach | 542 | 100 | 20 |
| 2 | Commercial area | 544 | 100 | 20 |
| 3 | Desert | 532 | 100 | 20 |
| 4 | Forest | 560 | 100 | 20 |
| 5 | Lake | 533 | 100 | 20 |
| 6 | Overpass | 541 | 100 | 20 |
| 7 | River | 493 | 100 | 20 |
| 8 | Tennis court | 530 | 100 | 20 |
| 9 | Wetland | 538 | 100 | 20 |
| Total no. of samples | | 5451 | 1000 | 99 |

**Table 8.3 Evaluation Metrics for Object Classification**

| Class Name | Precision | Recall | F1-Score |
|---|---|---|---|
| Airplane | 0.99 | 1.00 | 1.00 |
| Beach | 0.99 | 0.97 | 0.98 |
| Commercial area | 0.97 | 0.98 | 0.98 |
| Desert | 1.00 | 0.98 | 0.99 |
| Forest | 0.93 | 0.99 | 0.96 |
| Lake | 0.91 | 0.90 | 0.90 |
| Overpass | 0.98 | 0.99 | 0.99 |
| River | 0.88 | 0.88 | 0.88 |
| Tennis court | 0.97 | 0.98 | 0.98 |
| Wetland | 0.85 | 0.80 | 0.82 |
| Total/average | 0.95 | 0.95 | 0.95 |

Fig. 8.5 shows the accuracy performance of the VGG-16 Deep CNN model for both training and validation data sets. The accuracy of the model increases for each and every epoch. Similarly, loss of the model decreases for every epoch. The fluctuations in the accuracy from one epoch to another epoch depend on the regularization and optimizing techniques.

The generated confusion matrix is shown in Table 8.4. The correctly classified data items are placed in the diagonal of confusion matrix, and remaining missclassified data items are placed above and below the diagonal of the confusion matrix. We found that the error occurs when "wetland" classified as "forest" and "lake." Fig. 8.6 shows the ROC curve of the proposed model for object classification.

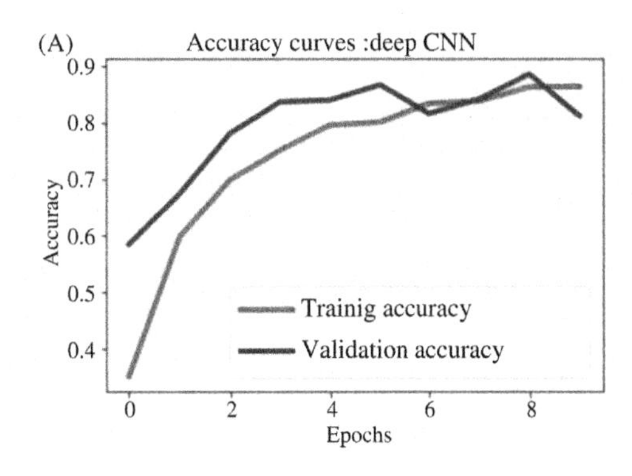


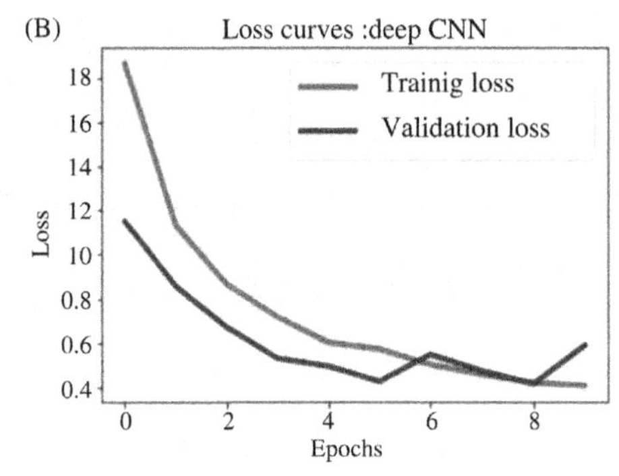


**FIGURE 8.5**

Train and validation accuracy for visual geometry group-16 deep convolutional neural network model. (A) Model accuracy for validation set. (B) Model loss for validation set.

**Table 8.4 Confusion Matrix for 10 Class Object Classification**

| Class Name | Airplane | Beach | Commercial Area | Desert | Forest | Lake | Overpass | River | Tennis Court | Wetland |
|---|---|---|---|---|---|---|---|---|---|---|
| Airplane | 100 | 0 | 0 | 0 | 0 | 0 | 1 | 0 | 0 | 0 |
| Beach | 0 | 97 | 0 | 0 | 0 | 0 | 0 | 1 | 0 | 2 |
| Commercial area | 1 | 0 | 98 | 0 | 0 | 0 | 0 | 1 | 0 | 0 |
| Desert | 0 | 0 | 0 | 98 | 1 | 0 | 0 | 1 | 0 | 0 |
| Forest | 0 | 0 | 0 | 0 | 99 | 0 | 0 | 0 | 0 | 1 |
| Lake | 0 | 0 | 0 | 0 | 0 | 90 | 0 | 2 | 0 | 8 |
| Overpass | 0 | 0 | 1 | 0 | 0 | 0 | 99 | 0 | 0 | 0 |
| River | 0 | 0 | 2 | 0 | 1 | 4 | 1 | 88 | 2 | 2 |
| Tennis court | 0 | 0 | 0 | 0 | 0 | 0 | 1 | 0 | 98 | 1 |
| Wetland | 0 | 1 | 0 | 0 | 6 | 5 | 0 | 7 | 1 | 80 |

## 8.5.1 CLASSIFICATION OF RESULTS FOR VARIOUS HYPERPARAMETERS

The VGG-16 model was trained by varying three hyperparameter activation function, dropout probability, and batch size. The results are shown in Tables 8.5–8.7 and its corresponding chart is shown in Figs. 8.7–8.9.

In this analysis, we found that the activation function "ReLU" has higher results than other two functions ELU and tanh. By using this activation function, the VGG-16 Deep CNN model achieves 0.947 accuracy result. The experiment by varying dropout probability gives better performance for 0.3. Also the efficiency of the VGG-16 deep CNN model was compared with different batch sizes such as 4, 8, 12, and 16 and found that the batch size 4 gave better results.

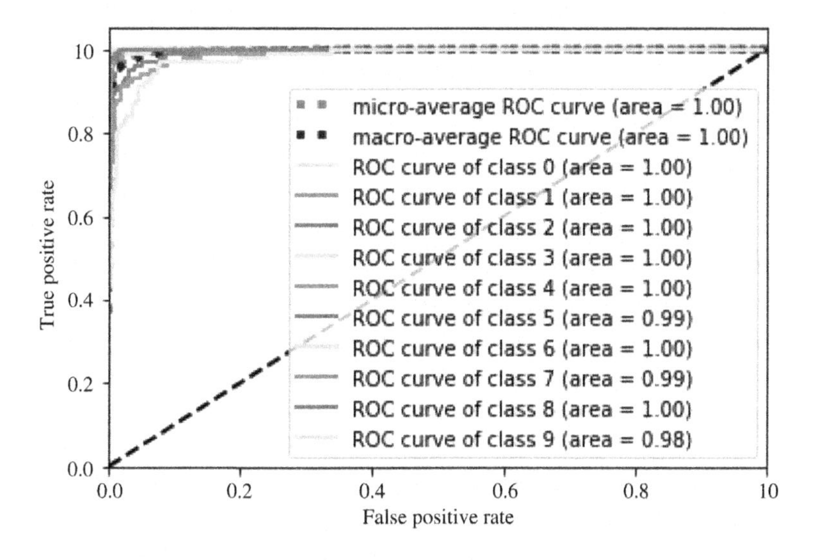


**FIGURE 8.6**

ROC curve of the proposed model for object classification.

**Table 8.5 Classification Performance of Various Activation Functions**

| Activation Function | Accuracy | Precision | Recall | F1-Score |
|---|---|---|---|---|
| ReLU | 94.7 | 95 | 95 | 95 |
| ELU | 90.5 | 91 | 91 | 90 |
| Tanh | 91.1 | 92 | 91 | 91 |

ReLU, *Rectified linear unit.*

**Table 8.6 Classification Performance of Various Batch sizes**

| Batch Size | Accuracy | Precision | Recall | F1-Score |
|---|---|---|---|---|
| Batch 4 | 94.7 | 95 | 95 | 95 |
| Batch 8 | 93.8 | 95 | 94 | 94 |
| Batch 12 | 93.5 | 94 | 94 | 94 |
| Batch 16 | 92.1 | 92 | 92 | 92 |

**Table 8.7 Classification Performance of Various Dropout Probabilities**

| Dropout | Accuracy | Precision | Recall | F1-Score |
|---|---|---|---|---|
| Dropout-0.2 | 90.5 | 92 | 91 | 91 |
| Dropout-0.3 | 94.7 | 95 | 95 | 95 |
| Dropout-0.4 | 91.2 | 91 | 90 | 90 |
| Dropout-0.5 | 92.1 | 92 | 92 | 92 |

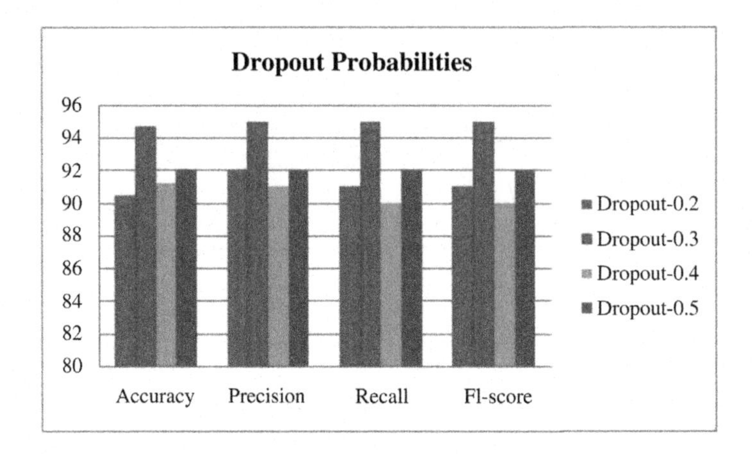


**FIGURE 8.7**

Classification performance chart of various dropout probabilities.

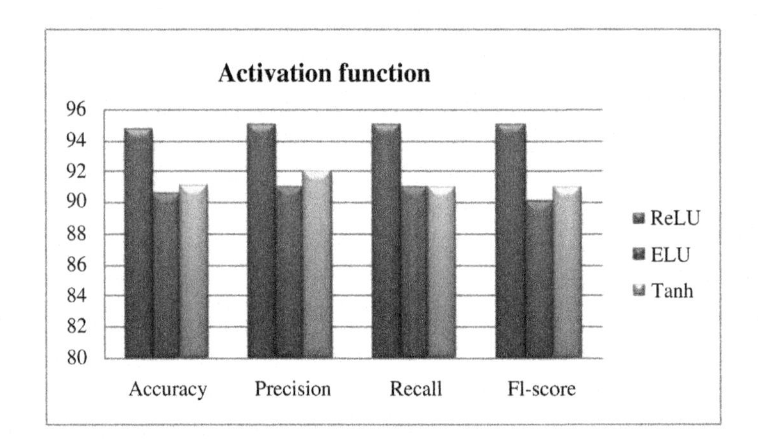


**FIGURE 8.8**

Classification performance chart of various activation functions.

## 8.6 CONCLUSION

In this paper, we have proposed object classification methods for remote sensing images using VGG-16 deep CNN. The deep CNN model acts as a feature extractor and classifier for the given training images as well as validation images. We have tuned the VGG-16 deep CNN with batch size 4, dropout-0.3, and activation function ReLU. With this mode, we have achieved 94.7% accuracy for 10 class remote sensing images, which is higher than other classification methods. In future, we have planned to implement the proposed work in GPU configuration for reducing the computational time.

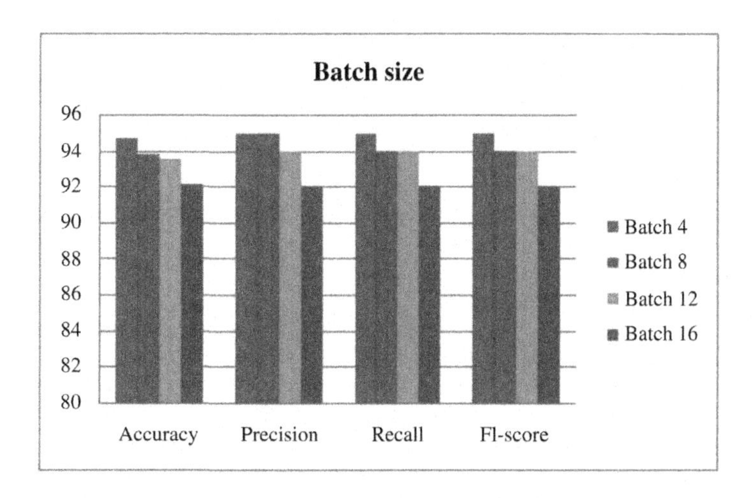


**FIGURE 8.9**

Classification performance chart of various batch sizes.

## REFERENCES


[1] P. Deepan, L.R. Sudha, Object detection in remote sensing images: a review, Int. J. Sci. Res. Computer Sci. Appl. Manag. Stud. (2018). ISSN 2319 – 1953.

[2] L. Janssen, H. Middelkoop, Knowledge based crop classification of a Landsat thematic mapper image, Int. J. Remote Sens. 13 (15) (1992) 2827–2837.

[3] T. Blaschke, Object based image analysis for remote sensing, ISPRS J. Photogramm. Remote Sens. 65 (2010) 2–16.

[4] L. Drăguţ, T. Blaschke, Automated classification of landform elements using object-based image analysis, Geomorphology 81 (2006) 330–344.

[5] G. Cheng, J. Han, X. Lu, Remote sensing image scene classification benchmark and state of the art, Proc. IEEE (2019) 1–19.

[6] L. Zhao, P. Tang, L. Huo, Feature significance-based multi bag-of-visual-words model for remote sensing image scene classification, J. Appl. Remote Sens. 10 (3) (2016). 035004.

[7] K. Qi, H. Wu, C. Shen, J. Gong, Land use scene classification in high-resolution remote sensing images using improved correlatons, IEEE Geosci. Remote Sens. Lett. 12 (12) (2015) 2403–2407.

[8] Y. Li, C. Tao, Y. Tan, K. Shang, J. Tian, Unsupervised multilayer feature learning for satellite image scene classification, IEEE Geosci. Remote. Sens. Lett. 13 (2) (2016).

[9] T.-H. Chan, K. Jia, S. Gao, J. Lu, Z. Zeng, Y. Ma, PCANet: a simple deep learning baseline for image classification? IEEE Trans. Image Process. 24 (12) (2015) 5017–5032.

[10] L. Zhang, L. Zhang, B. Du, Deep learning for remote sensing data: a technical tutorial on the state of the art, IEEE Geosci. Remote Sens. Mag. 4 (2) (2016) 22–40.

[11] K. Simonyan, A. Zisserman, Very deep convolutional networks for large-scale image recognition, in: Proc. Int. Conf. Learn. Represent, 2015, pp. 1–13.

[12] Y. Bazi, F. Melgani, Toward an optimal SVM classification system for hyperspectral remote sensing images, IEEE Trans. Geosci. Remote. Sens. 44 (11) (2006) 3374–3385.

[13] R. Archibald, G. Fann, Feature selection and classification of hyperspectral images with support vector machines, IEEE Geosci. Remote Sens. Lett. 4 (4) (2007) 674–677.

[14] T. Kavzoglua, P.M. Mather, The use of backpropagating artificial neural networks in land covers classification, Int. J. Remote Sens. 24 (23) (2003).

[15] M. Belgiu, L. Drăguţ, Random forest in remote sensing: a review of applications and future directions, ISPRS J. Photogramm. Remote Sens. 114 (2016) 24–31.

[16] M.M. Hayesa, S.N. Millera, M.A. Murphya, High-resolution landcover classification using random forest, Remote Sens. Lett. 5 (1) (2014) 112–121.

[17] A.M. Cheriyadat, Unsupervised feature learning for aerial scene classification, IEEE Trans. Geosci. Remote Sens. 52 (1) (2014) 439–451.

[18] S. Chaib, Y. Gu, H. Yao, An informative feature selection method based on sparse PCA for VHR scene classification, IEEE Geosci. Remote Sens. Lett. 13 (2) (2016) 147–151.

[19] F. Zhang, B. Du, L. Zhang, Scene classification via a gradient boosting random convolutional network framework, IEEE Trans. Geosci. Remote Sens. 54 (3) (2016) 1793–1802.

[20] G. Sheng, W. Yang, T. Xu, H. Sun, High-resolution satellite scene classification using a sparse coding based multiple feature combination, Int. J. Remote Sens. 33 (8) (2012) 2395–2412.

[21] Y. Zhong, F. Fei, L. Zhang, Large patch convolutional neural networks for the scene classification of high spatial resolution imagery, J. Appl. Remote Sens. 10 (2) (2016). 025006.

[22] M.C. Kruithof, H. Bouma, N.M. Fischer, K. Schutte, Object recognition using deep convolutional neural networks with complete transfer and partial frozen layers, in: Proc. Society of Photo-Optical Instrumentation Engineers, vol. 9995, 2016.

[23] Q. Zou, L. Ni, T. Zhang, Q. Wang, Deep learning based feature selection for remote sensing scene classification, IEEE Geosci. Remote Sens. Lett. 12 (11) 2015.

[24] N. Kussul, M. Lavreniuk, S. Skakun, A. Shelestov, Deep learning classification of land cover and crop types using remote sensing data, IEEE Geosci. Remote Sens. Lett. 14 (5) (2017).

[25] G. Hinton, N. Srivastava, A. Krizhevsky, I. Sutskever and R. Salakhutdinov, "Dropout: a simple way to prevent neural networks from overfitting", J. Mach. Learn. Res., pp. 1929–1958, 2014.

CHAPTER

# COMPRESSIVE SENSING-AIDED COLLISION AVOIDANCE SYSTEM

**P. Sreevidya[1], S. Veni[2], Vijay Rajeev[1] and P.U. Krishnanugrah[1]**

[1]*Federal Institute of Science and Technology, Angamaly, India* [2]*Amrita Institute of Science and Technology, Coimbatore, India*

## CHAPTER OUTLINE



## 9.1 INTRODUCTION

Since the inception of mathematical modeling of engineering systems, linear algebra has been used as a fundamental structure due to its abundant capacity for the presentation of geometry, planes, space, and rotation. This further aids the analysis of many natural phenomena. This decimation in complexity allows a logical analysis and efficient computations of the systems. Linear algebra accommodates a system of linear equations generally represented as $Ax = Y$, where $A$ represents all the $m$ coefficients, each from the set of $n$ equations. The very solvability coupled with the nature of solution is determined by the structure of the matrix $A$. But, the system becomes underdetermined, if $A$ is in an over-complete basis. An infinite set of dimensions can thus represent the same data set $x$, which disrupts the uniqueness of the solution. But the introduction of the criterion of sparsity in the vector space of data $x$ makes it complete and thus converges to a unique solution. It is from this very canny idea that evokes the practicality of compressive sensing (CS). Donoho and David [1] stated that a prior knowledge about a signal's sparsity can be used for reconstructing the signal

The Cognitive Approach in Cloud Computing and Internet of Things Technologies for Surveillance Tracking Systems.
DOI: https://doi.org/10.1016/B978-0-12-816385-6.00009-X

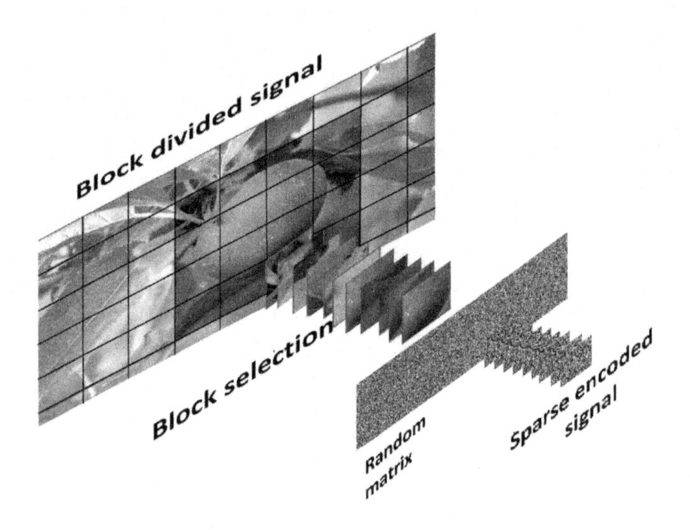


**FIGURE 9.1**

Block compressed sensing.

with far few samples as demanded by the sampling theorem. This ground breaking idea thus breaks free off the existing real-time systems from the constraint set by the sampling theorem. Fast, efficient, and inexpensive systems can thus be realized [2]. CS is a dump signal acquisition model, in the sense that it does not take account of neither the domain of sparsity of the signal nor the magnitude and the position within the signal. The randomness and the sporadic nature of signal acquisition brought about by CS add to their usefulness in critical system realizations. This chapter discusses one such critical application of CS in collision prevention system in automobiles. The framework consists of acquiring the digital image from the camera mounted on the vehicle. A Block Compressed Sensing (BCS) [3] approach is adopted here. This method reduces the computational complexity, memory, and time by sensing a randomly formed block divided pixel space (Fig. 9.1). In [3], it was studied that BCS produced comfortable performance in natural images. The online camera captures frames at 24 $\mathrm{fs}^{-1}$, where each of the frames is sensed by BCS. The acquired signal undergoes image processing to find a reference pixel space that infers to the information about the distance between vehicles. Section 9.2 gives a brief review over the existing CS literature. Sections 9.3 and 9.4 produce a detailed description and result analysis of the proposed system.

## 9.2 THEORETICAL BACKGROUND

According to Shannon, for recovering a continuous-time signal, it has to be sampled at a sampling frequency $f_s$ which is twice the maximum frequency component $f_m$ in the signal [4].

$$f_s = 2f_m \tag{9.1}$$

This theorem became the very fundamental element that bridged the analog and digital signals. In the existing signal processing paradigm, these uniformly collected $n$ samples are further compressed into $m$ numbers, where $m < n$ entities, discarding all the $n - m$ entries [2,5]. At the reception end, decompressions of these entities are carried out to represent the entire signal. When it comes to realization of a practical digital communication system, measuring $n$ samples, only to preserve "$m$" among them, turns out to be cumbersome. Furthermore, in the event of identifying all the "$m$" entities, the remaining $n - m$ entries were also to be operated, which accounted for a major loss in memory, power, and time. Furthermore, the peculiar character around the sampling theorem sticks to the dogma that denser the sampling, the fairer will become the reconstruction, which infringes with the flexibility of the realizable system. The current designing of signal acquisition system is constrained by the long-established tradition of the sampling theorem.

## 9.2.1 SPARSITY

It is observed that the signal that has a high time bandwidth product demands a high sampling rate. But in the wavelet and Gabor transform, they can be represented over a small frequency range [6,7]. These domains can typically provide an approximation over the sparsity of the given signal. Fig. 9.2 shows the 2-D wavelet decomposition of an image in the *Haar* wavelet in single level.

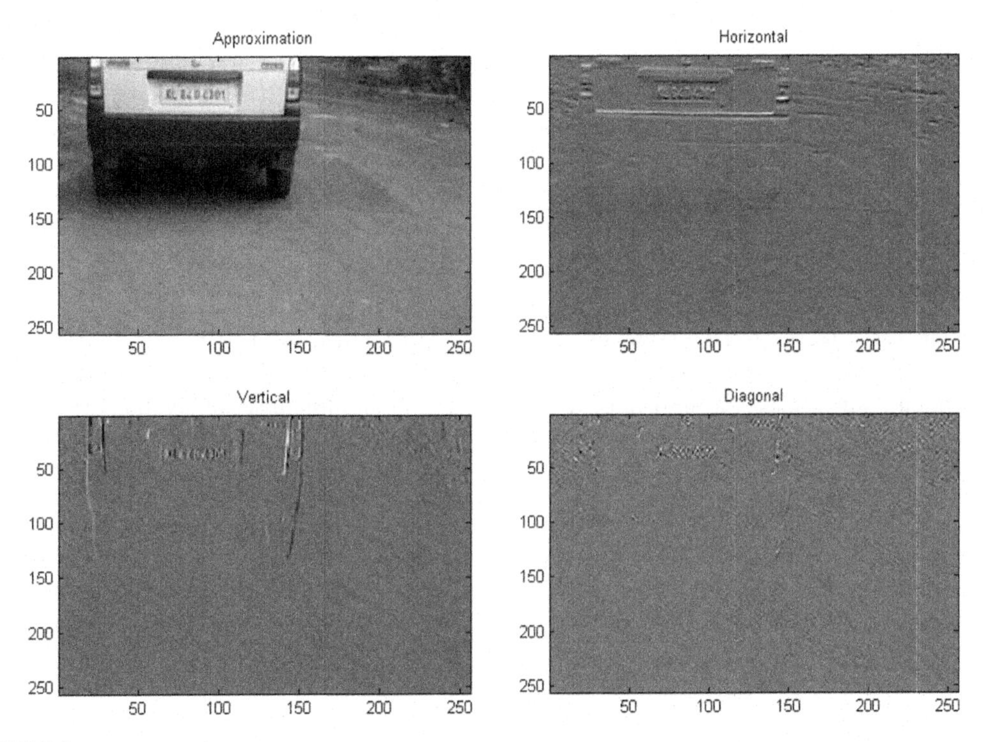


**FIGURE 9.2**

*Haar* wavelet decomposition.

These domains can represent the given signal $x$ with most of its coefficients having their absolute vales null [7]. This aids the judicious choice over the number of compression samples $m$ rather than a blind choice of a threshold value coupled with a tradeoff of quality. The sparsity of the signal is defined in the domain of choice as per the application. Under the current literature [6], the discovery of wavelet orthogonal bases lit a path for various transforms adaptable over a wide range of signals. As discussed in [7], these systems for sparse approximation can diverse from wavelet shearlets to dictionary learning algorithms, where the representational system is produced from a training set of signals [8]. The sparsity of the signal comes handy in the reduction of required measurements in the above discussed domains, for other way mammoth time-domain representations.

### 9.2.2 COMPRESSED SENSING PROBLEM STATEMENT

The CS problem statement can be introduced from a signal acquisition model for a signal $x$, where $(x_i)_{i=1}^{N} \in R^N$, where $x$ is a signal of the dimension $N$. It can be assumed that the signal $x$ itself is sparse, where most of the entries to $N$ dimensions are zero or negligible over $s$ larger absolute values [9]. In [10], a mathematical representation using the $l_0$ counting norm is used to evaluate the sparse nature of $x$, $|i{:}x_i \neq 0|$. But, it is very hard that the natural signals are expected to exhibit this character [11]. Hence, the existence of an orthonormal basis $\varphi$ over which $x$ is sparse is assumed.

$$x = \varphi c \tag{9.2}$$

where $c$ is the natural nonsparse vector of $x$. The orthonormal basis is often termed as a representational matrix [2]. The choice of a representational matrix is depended upon the application and nature of $x$. In a CS frame, random low dimensional measurements $m$ are taken over the signal by a measurement/sensing matrix $A$. $A$ is an $m$ dimensional vector space of $n$ vectors with $m$ much lesser than $n$. This system is coupled with a constraint that $x$ is sparse in some domain $\varphi$. The product of measurements over $x$ yields a measurement vector $Y$ in $m$ dimensions. The CS problem is stated as follows: recover $x$ from a prior knowledge of $A$ and $Y$.

$$Y = Ax \tag{9.3}$$

The above system represents a linear combination of $m$ dimensional vectors from $A$ with the corresponding entries in the signal $x$. The Eq. (9.2) represents an under-deterministic system of linear equations, which makes it unsolvable as it converges to an infinite set of solutions. The introduction of sparsity to the above equation will redesign it as follows:

$$Y = \psi c \tag{9.4}$$

where $\psi = A \times \varphi$. If the combined transform $\psi$ is a linear independent vector space, the problem is translated to an under-complete basis. The sparsity of $x$ forces $Y$ to be just $s$ linear combinations of $m$ dimensional vectors weighted by the corresponding sparse element's magnitude. This inhibits the contribution of all other $n - s$ vectors to $Y$. Fig. 9.3 depicts the CS as opposed to the traditional sampling technique.

### 9.2.3 RECOVERY

As $m > s$ and $\psi$ $(\psi = A \times \varphi)$ is linearly independent, the system takes unique redundant measurements on the signal that can be solved to retrieve the sparse solution. But, the nature of the

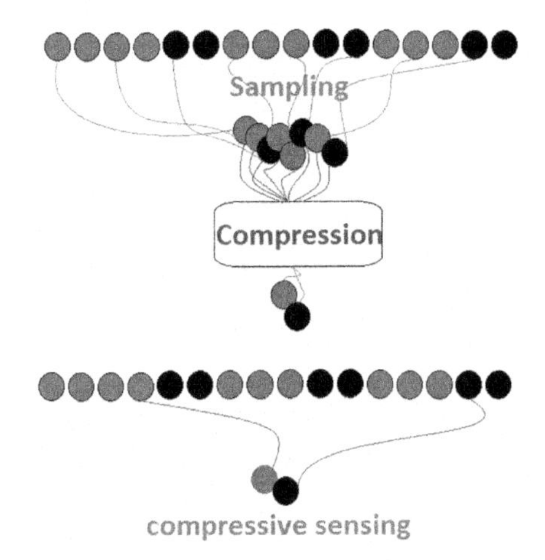


**FIGURE 9.3**

Traditional sampling versus compressed sensing.

measurement (active vectors) depends on the location of the sparse element in $x$. Solving for $x$ by trying out all the possible combinations of vectors makes the problem non-deterministic polynomial (NP)-hard [11]. Solution for $x$ can be founded out by the introduction of vector norms [1,10].

$$\min_x ||x||_0 \text{ subject to} ||Y - Ax||_2^2 = e \tag{9.5}$$

where $e$ is the expected error after recovery. As the signal is sparse, the obvious choice would be the pseudonorm $l_0$ [12]. But, this still does not lift the system from the burden of infinite combinatorial search and drives the problem NP-hard [11]. This calls for another suitable norm. In [13], it was proposed that the convex norm $l_1$ could replace $l_0$ in this minimization problem, which was termed *Basis Pursuit*:

$$\min_x ||x||_1 \text{ subject to} ||Y - Ax||_2^2 = e \tag{9.6}$$

The choice of $l_1$ over $l_2$ is due to the fact that it converges to the sparsest solution. This is evident from its geometric structure of $l_1$. Furthermore, $l_1$ is closest to $l_0$ in penalization, when $l_2$ exhibits a quadratic nature.

### 9.2.4 QUALITY MEASUREMENT

In digital image processing algorithms, quality assessment of a processed image is an integral part, as it is important to quantify the loss of information due to distortions and noises the image inherits. Under the current literature, there are various mathematical tools available for quality estimation [14]. In this chapter, two such popular techniques, namely peak signal-to-noise ratio (PSNR) and structural similarity index measure [SSIM], are used [15]. The computational easiness is one key factor that triggers its wide usage in most of the image processing algorithms [16]. From [16], for a reference image $f$ and its processed result $g$, both of the same dimension $M \times N$, PSNR is defined as follows:

$$\mathrm{PSNR}(f,g)=10\log_{10}\left(\frac{255^2}{\mathrm{MSE}(f,g)}\right)\frac{255^2}{\mathrm{MSE}(f,g)} \tag{9.7}$$

where

$$\mathrm{MSE}(f,g)=\frac{1}{MN}\sum_{(i=1)}^{M}\sum_{(j=1)}^{N}(f_{ij}-g_{ij})^2 \tag{9.8}$$

where MSE is the mean square error between the reference and reconstructed images. As the MSE approaches zero, the PSNR value approaches infinity. This implies that PSNR increases as the reconstruction produces an image of higher quality. However, PSNR lacks in the ability to examine the structural information of the image [17]. This surfaces the advantage of SSIM [18] over absolute error estimations like PSNR and MSE. SSIM takes account of the loss in correlation, luminance, and contrast masking associated with the reconstructed image. In [18], SSIM is defined as follows:

$$\mathrm{SSIM}(f,g)=l(f,g)c(f,g)s(f,g) \tag{9.9}$$

where

$$l(f,g)=\frac{2\mu_f\mu_g+C_1}{\mu_f^2+\mu_g^2+C_1} \tag{9.10}$$

$$c(f,g)=\frac{2\sigma_f\sigma_g+C_2}{\sigma_f^2\sigma_g^2+C_2} \tag{9.11}$$

$$s(f,g)=\frac{\sigma_{fg}+C_3}{\sigma_f\sigma_g+C_3} \tag{9.12}$$

The function *l(f,g)* is the luminance comparison function with the mean luminance of reference and processed images, $\mu_f$ and $\mu_g$. The second function *c(f,g)* compares the contrast between the two images with standard deviation $\sigma_f$ and $\sigma_g$. The structural function *s(f,g)* measures the structural comparison between the two images. $\sigma_{fg}$ is the covariance between the two images, and $C_1$, $C_2$, and $C_3$ are constants. From the studies conducted in [15,19], it was concluded that SSIM and PSNR were sensitive to addictive Gaussian noises. The proposed architecture introduces a collision prevention system in automobiles by taking a breaking action autonomously, overriding manual inputs. The digital image sensor mounted at the front of the vehicle captures frames at a fixed rate of 24 $\mathrm{fs}^{-1}$ (Fig. 9.4).

## 9.3 SYSTEM

The system, aided by BCS, acquires random limited measurements from these frames as the input. The acquired data undergo a series of operations (Sections 9.3.1, 9.3.2, and 9.3.3), comprising reconstruction and image processing converging to a single task of distance measurement between the object and the vehicle. The system is designed to produce a warning signal (visual, audio, or both). When the vehicle is in imminent collision region (between 2 and 1.5 m), the system measures the object distance by inferring the pixel density in the detected object. A fifth degree polynomial aids the distance prediction. The manual controls are overruled to apply the brakes, once the

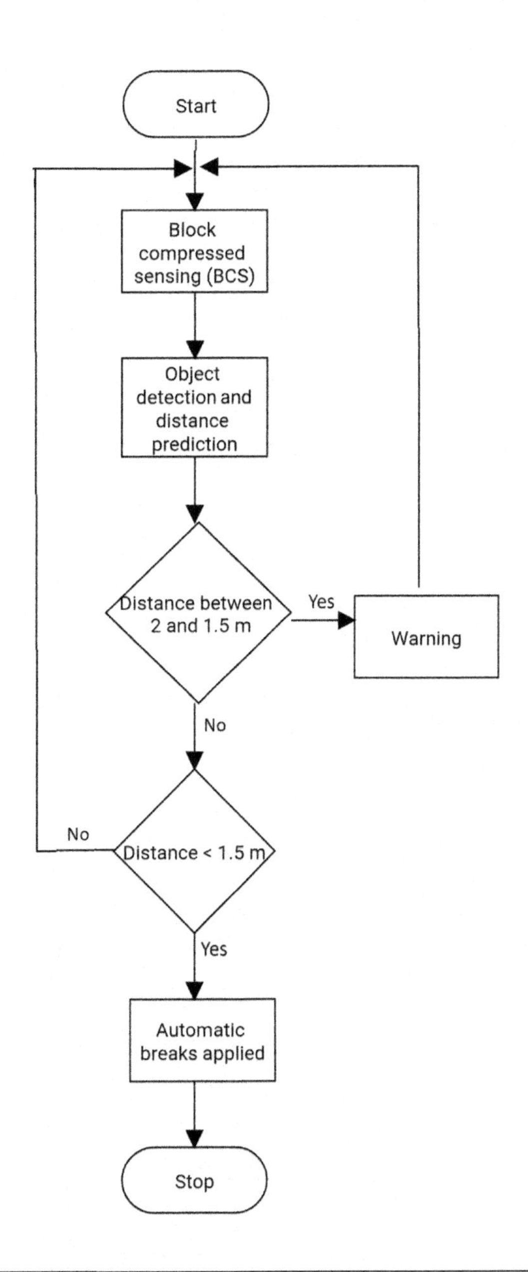


**FIGURE 9.4**

Block diagram of the proposed system.

distance crosses the critical threshold value (1.5 m). The predicted distance is compared with the threshold values, and appropriate actions are taken. Digital images being bulky add to the computational complexity and delay in the system. However, the system realization through BCS enhances the possibility of implementing the system in real time.

### 9.3.1 SIGNAL ACQUISITION

The signal acquisition is performed by mounting a digital camera in the front of the car. The proposed framework uses an SJCAM SJ4000 device with a Novatek 96650 chipset. The device contains AR0330, a 1/3-in. CMOS digital image sensor with the image resolution of $1080 \times 1920$, and an aperture of *f*/2.8. The camera streams an online video from which frames are captured at fixed intervals. In the designed framework, 24 frames are captured per second (Fig. 9.5).

The acquired frames undergo CS [3]. The system adopts a BCS technique. In this approach, the pixel space is divided into blocks of fixed dimensions. Fig. 9.6 shows a captured frame that is divided into blocks of $32 \times 32$. The pixel space can contain blocks of useless data that can be discarded. Only the blocks where the required information can be present are selected. Fig. 9.6 shows a captured frame of the dimensions of $1080 \times 1920$, of which the presence of desired information is distributed over a coordinate, [(554, 413) and (1460, 959)]. The BCS approach selectively takes the useful data from these pixel spaces. This further adds to the flexibility of the system. The BCS approach further reduces the memory requirements of the system.

The dimensions of the block are fixed to $32 \times 32$. Those pixel blocks that have the possibility of containing information are selected and reshaped as a vector of length $N$ ($N = 32 \times 32$) and stacked together to form a signal matrix (Fig. 9.6) of size $N \times L$, where $L$ represents the number of selected pixel blocks. Each of these transformed blocks later passed through a random measurement matrix (Fig. 9.7) to take fixed random measurements. This randomized measurement operations produce a matrix $Y$. This is then used to recover the image from the knowledge of measurement matrix and orthonormal basis $\varphi$. The image is assumed to be sparse in a wavelet basis. Hence, for the purpose of recovery, a *Haar* wavelet basis is chosen as $\varphi$.

### 9.3.2 IMAGE PROCESSING

Once the reconstruction of the frame is complete, the next step is to identify a reference segment that can be used as the reference for pixel variations. This reference point will serve as a mean to find the relative distance between the two vehicles. Number plates are identified as this common reference point, as they are all of almost the same fixed dimensions and locations. The processing is carried out in MATLAB 2014a in the experimental setup. The reconstructed image undergoes a

**FIGURE 9.5**

(A) Original frame. (B) Block divided frame.

**FIGURE 9.6**

(A) Identified blocks containing information. (B) Area corresponding to selected blocks.

**FIGURE 9.7**

Signal transformation.

series of image processing operations here. The MATLAB R2014a software [20] is adopted, as it ensures better system prototyping [21]. The reconstructed images are subjected to morphological operations [22] to locate spatial pixel objects and to segment the desired dimensions at expected locations. These operations produce a structure of pixel objects as in figure 9.9 (A) and 9.10 (A). For the purpose of plate detection, all pixel objects less than the threshold value are neglected. From the analysis of various sets of operated images, this threshold value was set to 3000 pixels. Out of these, objects with a major axis orientation between $-20$ and $+20$ are filtered. In practice, it was observed that these selected features were fruitful in mapping just the area corresponding to the number plate, as it showed a positive result over all the frames (Fig. 9.8).

| | 1 | 2 | 3 | 4 | 5 | 6 | 7 | 8 | 9 | 10 | 11 | 12 |
|---|---|---|---|---|---|---|---|---|---|---|---|---|
| 1 | 0.5377 | -0.4807 | -0.3982 | 0.9195 | -1.2717 | -0.3273 | 1.4139 | 0.6541 | -0.5487 | -2.5611 | -1.2128 | 0.8688 |
| 2 | 1.8339 | 0.8368 | -0.0855 | 0.8522 | 0.2726 | 0.7421 | 0.5907 | -0.3410 | -2.2004 | 0.3042 | -0.3812 | 0.1358 |
| 3 | -2.2588 | 2.5383 | 2.3733 | 0.6571 | 0.3175 | -1.0057 | 0.7800 | -0.8433 | -1.2631 | -1.0833 | 1.3227 | 0.5305 |
| 4 | 0.8622 | -1.3233 | -0.4739 | -0.7535 | 1.4984 | 0.1194 | -0.6572 | -0.7299 | -0.4164 | 0.6348 | 0.3853 | 0.0232 |
| 5 | 0.3188 | 0.1283 | 0.9463 | -1.3385 | -1.5538 | 0.2004 | 0.1318 | -1.2966 | 0.3269 | 0.6029 | -0.3618 | -0.4483 |
| 6 | -1.3077 | -1.4424 | 0.8182 | 0.6522 | -0.3543 | 1.5739 | -1.1224 | -0.8237 | -0.8441 | 1.0182 | -0.4264 | -0.6585 |
| 7 | -0.4336 | 1.3025 | 1.5890 | 1.4472 | 0.4342 | 0.5386 | 0.2398 | 1.1819 | -1.2142 | 0.6461 | 0.3871 | -0.7852 |
| 8 | 0.3426 | 1.4099 | 0.5260 | -1.2901 | -0.1015 | 0.0331 | 1.1422 | -0.5983 | -1.1062 | 0.7714 | 0.9028 | -0.8470 |
| 9 | 3.5784 | -1.6625 | -2.4652 | -2.2082 | 2.0288 | -1.3417 | -1.1009 | -0.0596 | 0.7967 | -0.2778 | 0.7178 | 0.0109 |
| 10 | 2.7694 | 1.9437 | -0.8525 | 1.4361 | -0.3672 | -1.4709 | -0.2089 | 1.5574 | 0.2774 | 2.2408 | 1.7361 | -1.5884 |
| 11 | -1.3499 | -1.0847 | 0.5117 | -0.0617 | -2.3638 | 0.5906 | -1.4661 | -0.4987 | 0.1436 | 0.2678 | -1.0088 | -0.1280 |
| 12 | 3.0349 | 0.2268 | 0.2578 | 1.1778 | 0.7299 | -0.7002 | 0.6063 | -0.2014 | 0.7804 | 0.3996 | -0.1800 | -1.5132 |
| 13 | 0.7254 | 1.0989 | 1.9624 | 0.9855 | -1.3990 | -0.2412 | 1.2308 | 0.6913 | 0.8574 | 1.3195 | -2.0265 | -0.2684 |
| 14 | -0.0631 | 0.1472 | 1.4063 | -1.2186 | 1.3149 | -0.8694 | 1.1435 | 0.1994 | -0.6256 | -1.0434 | 0.4089 | -0.9203 |
| 15 | 0.7147 | 2.2957 | 0.4968 | -0.4319 | 0.4038 | 0.1816 | -1.0257 | 0.4626 | -1.7229 | 3.0315 | 0.7343 | 0.1458 |
| 16 | -0.2050 | 2.7526 | 0.0828 | -0.8349 | -0.3442 | 0.2279 | -0.7090 | -0.3784 | -1.4578 | -1.2161 | 0.0820 | 0.9731 |
| 17 | -0.1241 | 0.1383 | -1.5485 | 0.1797 | -0.9801 | 1.7447 | -0.1022 | -0.1769 | 0.9008 | 1.4270 | -1.6865 | -1.6800 |
| 18 | 1.4897 | -1.9071 | 1.8632 | 1.1663 | 1.8573 | 1.8179 | -0.3951 | 1.6199 | 0.4708 | 1.3355 | -2.0617 | -0.7486 |
| 19 | 1.4090 | -0.3650 | 0.1340 | 0.0560 | 0.3095 | 1.0616 | -0.5276 | 2.0215 | -1.3936 | -0.6676 | 1.8209 | -0.7490 |
| 20 | 1.4172 | -0.8481 | -1.5460 | -2.1000 | -0.4891 | 1.0840 | -2.5245 | -0.0344 | -0.4147 | -1.9715 | -1.2491 | -0.4607 |

**FIGURE 9.8**

MATLAB-generated random measurement matrix.

As a preprocessing step for morphological operations, the recovered gray-scale image is transformed into a binary format of a supported data type. After a series of dilation and erosion operations [23,24] using appropriate structuring elements, the returned image contained a spatially distributed pixel objects of varying features. On these objects, the above discussed feature selection procedure was executed, which aided to spatially allocate just the number plate. This operation when repeated over a different set of frames produced a positive result (Fig. 9.9). This further strengthened the choice over features used for plate detection and its application into the system (Fig. 9.10).

### 9.3.3 ANALYSIS

Once the number plate is identified, it is taken for distance measurement. The idea behind this is that, when an object comes closer to an image acquisition point, the pixel space accommodated by the object within the frame increases. This increment in pixels is directly proportional to the distance between the image acquisition point and the object. In order to establish a relation of pixel variation with distance, manual measurements of length between the acquisition point and the object was carried out. From the distance labeled point, four sample measurements were taken, and an average of these sample values was used as the final pixel number. The distance labeled images were later examined to find their accommodating pixels.

Table 9.1 calibrates the relation between the pixel number and the distance in centimeters. Columns 2–5 display the recorded sample measurement, column 6 gives the average pixels, and column 1 shows the labeled distance. Inferring the data from Table 9.1, the characteristics

(A)

| Fields | Area | BoundingBox | Orientation |
|---|---|---|---|
| 1 | 841 | [0.5000,0.5000,0.... | [] |
| 2 | 4128 | [0.5000,0.5000,0.... | [] |
| 3 | 2304 | [0.5000,0.5000,0.... | [] |
| 4 | 1098 | [1.5000,0.5000,0.... | [] |
| 5 | 591 | [0.5000,0.5000,0.... | [] |
| 6 | 159 | [1.5000,1.5000,0.... | [] |
| 7 | 78 | [9.5000,1.5000,0.... | [] |
| 8 | 24 | [1.5000,25.5000,0... | [] |
| 9 | 18 | [18.5000,3.5000,0... | [] |
| 10 | 9 | [99.5000,2.5000,0... | [] |
| 11 | 45 | [19.5000,5.5000,0... | [] |
| 12 | 42 | [1.5000,10.5000,0... | [] |
| 13 | 138 | [4.5000,1.5000,0.... | [] |
| 14 | 255 | [0.5000,1.5000,0.... | [] |
| 15 | 462 | [1.5000,1.5000,0.... | [] |
| 16 | 477 | [1.5000,0.5000,0.... | [] |
| 17 | 513 | [3.5000,0.5000,0.... | [] |
| 18 | 498 | [0.5000,0.5000,0.... | [] |
| 19 | 456 | [4.5000,1.5000,0.... | [] |
| 20 | 219 | [9.5000,1.5000,0.... | [] |

(B)

| Fields | Area | BoundingBox | Orientation |
|---|---|---|---|
| 1 | 0 | 0 | [] |
| 2 | 4128 | [0.5000,0.5000,0.... | [] |
| 3 | 0 | 0 | [] |
| 4 | 0 | 0 | [] |
| 5 | 0 | 0 | [] |
| 6 | 0 | 0 | [] |
| 7 | 0 | 0 | [] |
| 8 | 0 | 0 | [] |
| 9 | 0 | 0 | [] |
| 10 | 0 | 0 | [] |
| 11 | 0 | 0 | [] |
| 12 | 0 | 0 | [] |
| 13 | 0 | 0 | [] |
| 14 | 0 | 0 | [] |
| 15 | 0 | 0 | [] |
| 16 | 0 | 0 | [] |
| 17 | 0 | 0 | [] |
| 18 | 0 | 0 | [] |
| 19 | 0 | 0 | [] |
| 20 | 0 | 0 | [] |

(C)

| Fields | Area | BoundingBox | Orientation |
|---|---|---|---|
| 237 | 170 | [7.5000,0.5000,89... | 10.1727 |
| 238 | 258 | [2.5000,1.5000,90... | 12.0526 |
| 239 | 288 | [35.5000,0.5000,8... | 10.9542 |
| 240 | 335 | [4.5000,0.5000,90... | 10.7886 |
| 241 | 367 | [0.5000,0.5000,90... | 9.6412 |
| 242 | 435 | [1.5000,0.5000,90... | 11.5972 |
| 243 | 534 | [1.5000,0.5000,90... | 9.9912 |
| 244 | 627 | [6.5000,0.5000,90... | 10.7566 |
| 245 | 656 | [3.5000,1.5000,90... | 10.0712 |
| 246 | 769 | [27.5000,0.5000,8... | 10.3902 |
| 247 | 758 | [36.5000,0.5000,8... | 11.2590 |
| 248 | 874 | [5.5000,0.5000,89... | 10.7726 |
| 249 | 973 | [1.5000,0.5000,90... | 10.8098 |
| 250 | 1073 | [6.5000,0.5000,90... | 10.7033 |
| 251 | 1262 | [0.5000,0.5000,90... | 10.9081 |
| 252 | 1993 | [2.5000,0.5000,90... | 8.1196 |
| 253 | 2798 | [0.5000,0.5000,90... | 11.4291 |
| 254 | 775 | [0.5000,0.5000,90... | 6.7824 |
| 255 | 30328 | [0.5000,0.5000,90... | 10.6497 |

(D)

| Fields | Area | BoundingBox | Orientation |
|---|---|---|---|
| 237 | 0 | 0 | 10.1727 |
| 238 | 0 | 0 | 12.0526 |
| 239 | 0 | 0 | 10.9542 |
| 240 | 0 | 0 | 10.7886 |
| 241 | 0 | 0 | 9.6412 |
| 242 | 0 | 0 | 11.5972 |
| 243 | 0 | 0 | 9.9912 |
| 244 | 0 | 0 | 10.7566 |
| 245 | 0 | 0 | 10.0712 |
| 246 | 0 | 0 | 10.3902 |
| 247 | 0 | 0 | 11.2590 |
| 248 | 0 | 0 | 10.7726 |
| 249 | 0 | 0 | 10.8098 |
| 250 | 0 | 0 | 10.7033 |
| 251 | 0 | 0 | 10.9081 |
| 252 | 0 | 0 | 8.1196 |
| 253 | 0 | 0 | 11.4291 |
| 254 | 0 | 0 | 6.7824 |
| 255 | 30328 | [0.5000,0.5000,90... | 10.6497 |

**FIGURE 9.9**

(A), (C), (E), and (G) Generated object structure. (B), (D), (F), and (H) Selection of desired object after feature extraction.

(E)

| Fields | Area | BoundingBox | Orientation |
|---|---|---|---|
| 32 | 11 | [38.5000,90.5000,... | -13.2825 |
| 33 | 1 | [40.5000,44.5000,... | 0 |
| 34 | 2 | [40.5000,83.5000,... | 0 |
| 35 | 2 | [41.5000,85.5000,... | 0 |
| 36 | 1 | [41.5000,87.5000,... | 0 |
| 37 | 1 | [44.5000,84.5000,... | 0 |
| 38 | 3006 | [45.5000,0.5000,1... | 33.8730 |
| 39 | 1 | [65.5000,78.5000,... | 0 |
| 40 | 1 | [69.5000,77.5000,... | 0 |
| 41 | 1 | [91.5000,77.5000,... | 0 |
| 42 | 273 | [123.5000,115.50... | -89.4311 |
| 43 | 81 | [206.5000,60.500... | -1.3751 |
| 44 | 4859 | [239.5000,2.5000,... | -1.3389 |
| 45 | 9 | [269.5000,58.500... | 0 |
| 46 | 51 | [280.5000,59.500... | -1.7816 |
| 47 | 1054 | [342.5000,42.500... | 44.4659 |
| 48 | 23 | [377.5000,71.500... | 0 |
| 49 | 1 | [446.5000,45.500... | 0 |
| 50 | 1 | [469.5000,51.500... | 0 |
| 51 | 1 | [459.5000,53.500... | 0 |

(F)

| Fields | Area | BoundingBox | Orientation |
|---|---|---|---|
| 32 | 0 | 0 | -13.2825 |
| 33 | 0 | 0 | 0 |
| 34 | 0 | 0 | 0 |
| 35 | 0 | 0 | 0 |
| 36 | 0 | 0 | 0 |
| 37 | 0 | 0 | 0 |
| 38 | 0 | 0 | 33.8730 |
| 39 | 0 | 0 | 0 |
| 40 | 0 | 0 | 0 |
| 41 | 0 | 0 | 0 |
| 42 | 0 | 0 | -89.4311 |
| 43 | 0 | 0 | -1.3751 |
| 44 | 4859 | [239.5000,2.5000,... | -1.3389 |
| 45 | 0 | 0 | 0 |
| 46 | 0 | 0 | -1.7816 |
| 47 | 0 | 0 | -44.4659 |
| 48 | 0 | 0 | 0 |
| 49 | 0 | 0 | 0 |
| 50 | 0 | 0 | 0 |
| 51 | 0 | 0 | 0 |

(G)

| Fields | Area | BoundingBox | Orientation |
|---|---|---|---|
| 1 | 401 | [0.5000,0.5000,0.... | 0 |
| 2 | 4914 | [0.5000,0.5000,0.... | 15.3200 |
| 3 | 2952 | [0.5000,0.5000,0.... | -2.6400 |
| 4 | 1368 | [1.5000,0.5000,0.... | -1.2500 |
| 5 | 741 | [2.5000,0.5000,0.... | 10.3500 |
| 6 | 279 | [1.5000,1.5000,0.... | 4.3200 |
| 7 | 108 | [2.5000,5.5000,0.... | 0 |
| 8 | 30 | [25.5000,1.5000,0... | 0 |
| 9 | 9 | [94.5000,10.5000,... | 0 |
| 10 | 15 | [22.5000,10.5000,... | 0 |
| 11 | 48 | [4.5000,1.5000,0.... | 10.2500 |
| 12 | 78 | [18.5000,2.5000,0... | 25.3200 |
| 13 | 189 | [1.5000,1.5000,0.... | 0 |
| 14 | 366 | [1.5000,1.5000,0.... | 0 |
| 15 | 561 | [0.5000,0.5000,0.... | 0 |
| 16 | 741 | [0.5000,0.5000,0.... | 0 |
| 17 | 633 | [1.5000,1.5000,0.... | 0 |
| 18 | 756 | [4.5000,1.5000,0.... | 0 |
| 19 | 507 | [0.5000,1.5000,0.... | 0 |
| 20 | 330 | [1.5000,1.5000,0.... | 0 |

(H)

| Fields | Area | BoundingBox | Orientation |
|---|---|---|---|
| 1 | 401 | 0 | 0 |
| 2 | 4914 | [0.5000,0.5000,0.... | 15.3200 |
| 3 | 0 | 0 | -2.6400 |
| 4 | 0 | 0 | -1.2500 |
| 5 | 0 | 0 | 10.3500 |
| 6 | 0 | 0 | 4.3200 |
| 7 | 0 | 0 | 0 |
| 8 | 0 | 0 | 0 |
| 9 | 0 | 0 | 0 |
| 10 | 0 | 0 | 0 |
| 11 | 0 | 0 | 10.2500 |
| 12 | 0 | 0 | 25.3200 |
| 13 | 0 | 0 | 0 |
| 14 | 0 | 0 | 0 |
| 15 | 0 | 0 | 0 |
| 16 | 0 | 0 | 0 |
| 17 | 0 | 0 | 0 |
| 18 | 0 | 0 | 0 |
| 19 | 0 | 0 | 0 |
| 20 | 0 | 0 | 0 |

**FIGURE 9.9**

(Continued).

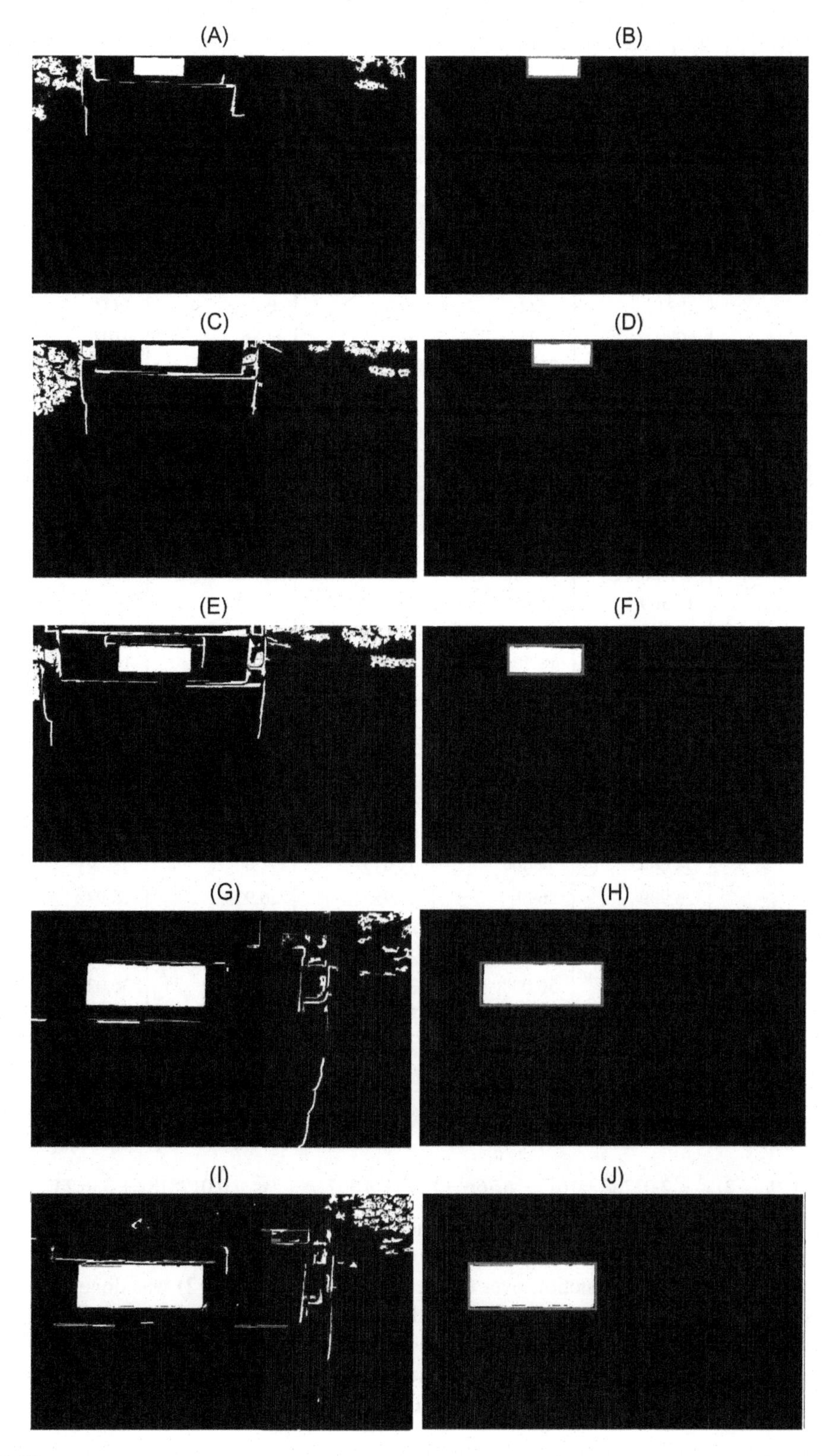


**FIGURE 9.10**

(A), (C), (E), (G), and (I) Frames after morphological operations. (B), (D), (F), (H) and (J) Result after feature extraction.

**Table 9.1 Distance–Pixel Characteristics.**

| | Pixel Values | | | | |
|---|---|---|---|---|---|
| **Distance (cm)** | **$S_1$** | **$S_2$** | **$S_3$** | **$S_4$** | **Average Pixel Value** |
| 280 | 4831 | 4844 | 4826 | 4833 | 4834 |
| 270 | 6090 | 6121 | 6131 | 6086 | 6107 |
| 260 | 7120 | 7180 | 7185 | 7135 | 7155 |
| 250 | 8048 | 8052 | 7991 | 7949 | 8010 |
| 240 | 9225 | 9198 | 9230 | 9215 | 9217 |
| 230 | 10,288 | 10,254 | 10,271 | 10,299 | 10,278 |
| 220 | 11,159 | 11,100 | 11,121 | 11,072 | 11,113 |
| 210 | 12,799 | 12,775 | 12,762 | 12,744 | 12,770 |
| 200 | 13,951 | 13,908 | 13,948 | 13,949 | 13,939 |
| 190 | 15,028 | 15,048 | 15,051 | 15,021 | 15,037 |
| 180 | 16,309 | 16,351 | 16,349 | 16,323 | 16,333 |
| 170 | 17,340 | 17,358 | 17,388 | 17,358 | 17,361 |
| 160 | 18,499 | 18,454 | 18,477 | 18,450 | 18,470 |
| 150 | 19,502 | 19,555 | 19,542 | 19,541 | 19,535 |
| 140 | 20,071 | 20,101 | 20,105 | 20,105 | 20,090 |
| 130 | 21,674 | 21,703 | 21,705 | 21,705 | 21,691 |
| 120 | 22,560 | 22,531 | 22,573 | 225,998 | 22,549 |
| 110 | 23,655 | 23,685 | 23,672 | 23,672 | 23,668 |
| 100 | 24,791 | 24,791 | 24,746 | 24,816 | 24,776 |
| 90 | 25,999 | 25,999 | 25,972 | 25,998 | 25,981 |
| 80 | 26,999 | 2699 | 26,955 | 26,973 | 26,972 |
| 70 | 27,751 | 27,751 | 27,749 | 27,872 | 27,767 |
| 60 | 28,922 | 28,922 | 28,891 | 28,896 | 28,906 |
| 50 | 30,008 | 30,028 | 30,038 | 30,010 | 30,021 |
| 40 | 31,501 | 31,548 | 31,544 | 31,499 | 31,523 |

between the number of pixels and the distance were plotted, and a suitable fifth degree equation was found to fit the curve [25]. Fig. 9.11 shows the data line along with the curve as estimated by Eq. (9.13)

$$Y = 6.119e - 21x^5 - 2.0276e - 16x^4 - 6.76e - 12x^3 + 3.3072e - 7x^2 - 0.012818x + 336.24 \quad (9.13)$$

where $Y$ is the distance in centimeters and $X$ is the number of pixels. The area returned from the detected pixel object is substituted to the above equation to find the distance.

Validation of the estimated function over the actual distance (Table 9.2) was done. The distance as estimated by the function (column 2) is observed over the actual distance (column 1). The differences in both of these measurements are founded out as errors (column 3). Fig. 9.12 shows a graphical representation of Table 9.2, with the actual distance in the X-axis and the estimated distance in the Y-axis. The mean absolute percentage error (MAPE) [26] for the overall system was calculated as follows:

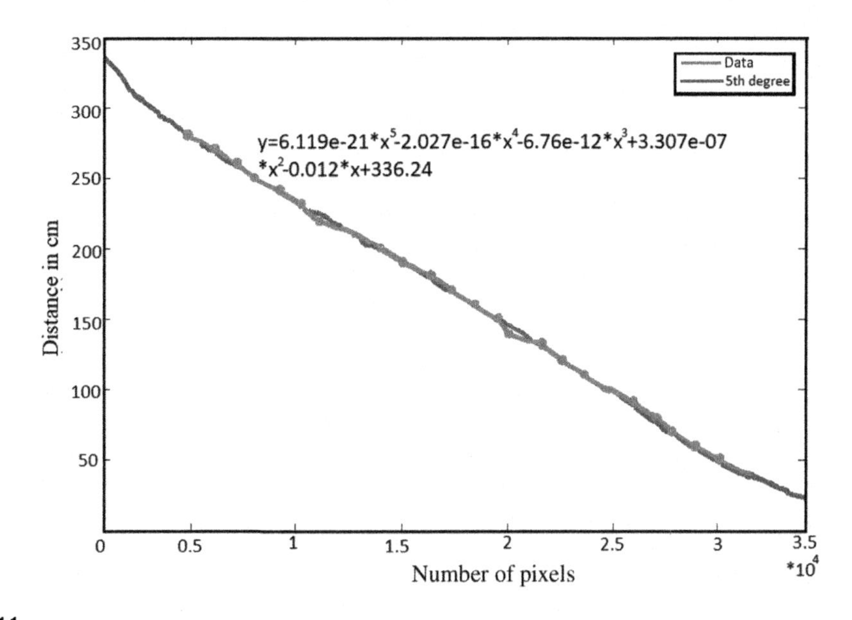


**FIGURE 9.11**

Curve fitting with the fifth degree polynomial.

**Table 9.2 Error.**

| S. No. | Error | | |
|---|---|---|---|
| | Actual Distance in Centimeters | Estimated Distance in Centimeters | %Error |
| 1 | 280 | 281 | 0.3571 |
| 2 | 270 | 269 | 0.3571 |
| 3 | 260 | 259 | −0.3571 |
| 4 | 250 | 251 | −0.3571 |
| 5 | 240 | 240 | 0 |
| 6 | 230 | 231 | 0.3571 |
| 7 | 220 | 223 | 1.0714 |
| 8 | 210 | 209 | −0.3571 |
| 9 | 200 | 199 | −0.3571 |
| 10 | 190 | 190 | 0 |
| 11 | 180 | 178 | −0.714 |
| 12 | 170 | 169 | −0.3571 |
| 13 | 160 | 159 | −0.3571 |
| 14 | 150 | 150 | 0 |
| 15 | 140 | 144 | 1.45286 |
| 16 | 130 | 129 | −0.3571 |
| 17 | 120 | 121 | 0.3571 |
| 18 | 110 | 110 | 0 |
| 19 | 100 | 99.6 | −0.1429 |
| 20 | 90 | 87.9 | −0.7500 |
| 21 | 80 | 78.5 | −0.5357 |
| 22 | 70 | 71 | 0.3571 |
| 23 | 60 | 60.7 | 0.2500 |
| 24 | 50 | 51.1 | −0.3929 |
| 25 | 40 | 39.3 | −0.2500 |

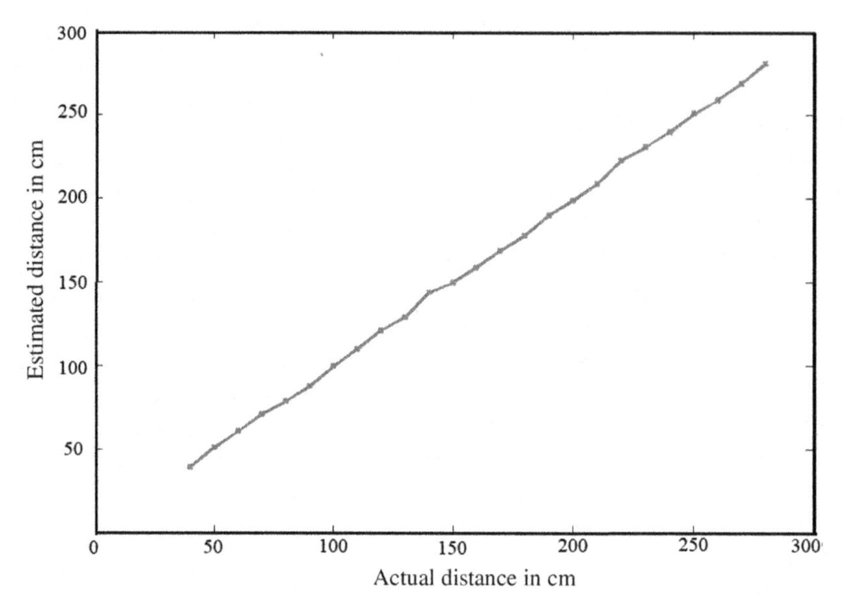


**FIGURE 9.12**

Error curve.

$$M = 1n\sum_{n}^{t=1}\left|\frac{A_t - E_t}{A_t}\right| \tag{9.14}$$

where $A_t$ is the actual value and $E_t$ is the estimated value at the $t$th measurement. $n$ represents the total number of measurements taken. For the proposed system, the MAPE was found to be 0.00392828.

## 9.4 RESULT

Frames were captured from the live streaming video at a rate of 24 $\text{fs}^{-1}$. From these frames, $M$ samples are taken, which are determined by the subrate $r$. The relation between the subrate and the number of samples are given by

$$M = r \times \text{block} - \text{size} \times \text{block} - \text{size} \tag{9.15}$$

where the block size is fixed to 32. The quality assessment of the reconstructed image is carried out in MATLAB. As discussed in Section 9.2, quality measurement, PSNR, and SSIM metrics are chosen. The MATLAB functions *psnr()* and *ssim()* are used over the reconstructed image to find PSNR and SSIM at a different subrate (Fig. 9.13). The reconstructed images are processed to check for the distance of the vehicle in front. By performing morphological operations, the number plate of the car is acquired, and this number plate is used as a reference to check the distance with the car in front by checking the pixel density inside the number plate. The distance is compared with a

| Recovered signal | | | | |
|---|---|---|---|---|
| subrate | 0.8 | 0.6 | 0.4 | 0.2 |
| samples *M* | 819 | 614 | 409 | 204 |
| PSNR | 53.61 | 52.35 | 49.59 | 43.10 |
| SSIM | 0.9272 | 0.9137 | 0.8858 | 0.7914 |

**FIGURE 9.13**

Peak signal-to-noise ratio and structural similarity index measure for the recovered image.

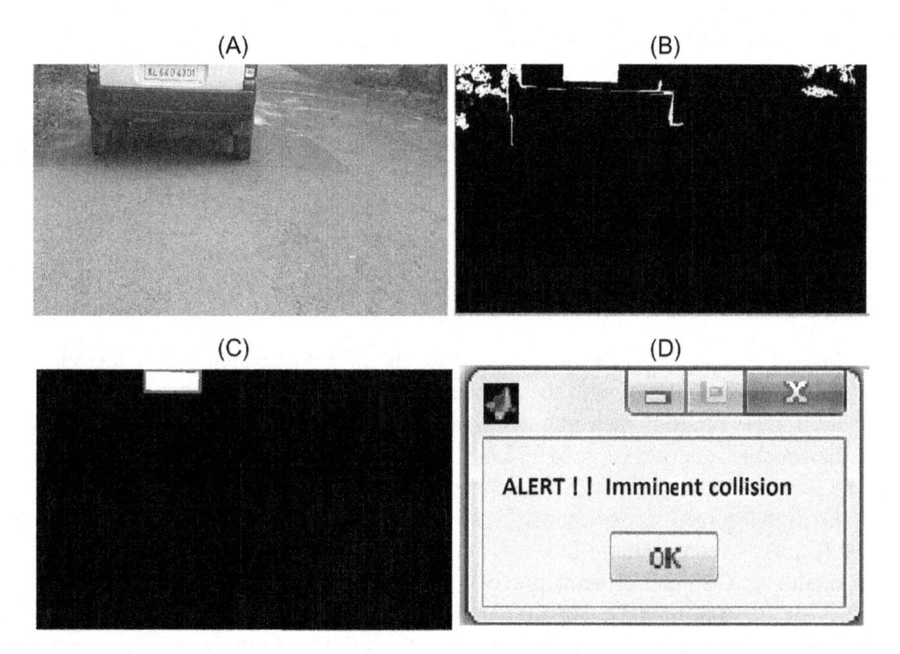


**FIGURE 9.14**

Vehicle at an imminent collision range. (A) Block compressed sensing sensed frame. (B) Identified objects. (C) Plate detection. (D) Activated warning signal.

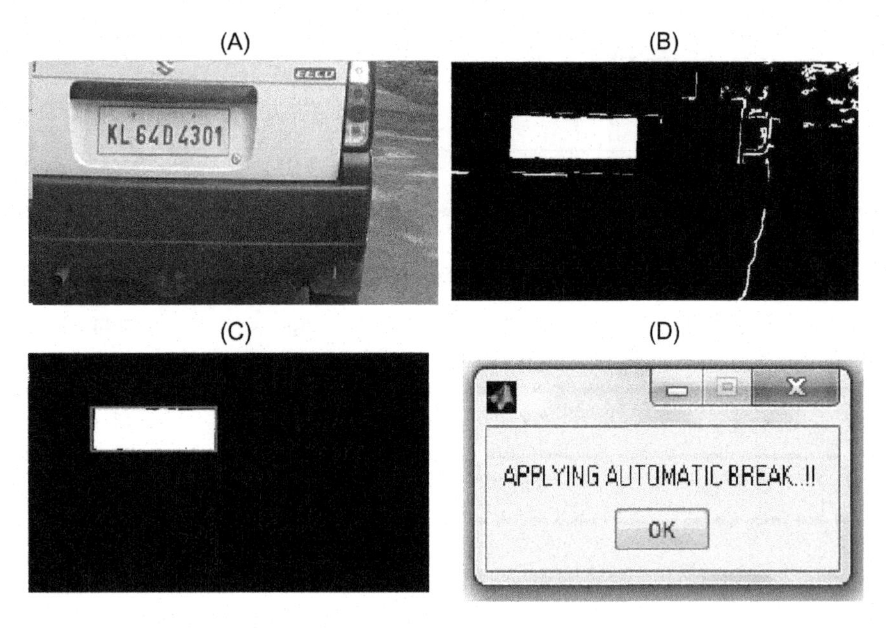


**FIGURE 9.15**

System generated result for the vehicle closer than the threshold value. (A) Block compressed sensing sensed frame. (B) Identified objects. (C) Plate detection. (D) Application of automatic braking.

tolerant range between 2 and 1.5 m. When the result of the comparison turns true, a warning signal is generated (Fig. 9.14). As the vehicle crosses the critical threshold value of 1.5 m, automatic breaking is applied (Fig. 9.15).

## 9.5 CONCLUSION

The system was successful in recovering the entire signal from a handful of measurements $M$, which is very small in comparison with the number of measurements as demanded by the Shannon sampling theorem. The random measurements were taken by using a randomly formed matrix, formed from the *randn()* function from MATLAB. These measurements guaranteed a perfect reconstruction of the original signal with a PSNR depended on the selected subrate. This further confirms the fact that the "randomized measurements" can encode the sparse position and magnitude. The validity of the $l_1$ norm over the signal recovery problems is reinforced. We were able to implement the software model of CS for collision prevention system, which can be much faster in signal acquisition than the existing models. The system was simulated in an Intel Core i5 7500 Processor. It was observed that the signal recovery time varied from 5 to 25 seconds depending on the selected subrate ranging from 0.2 to 0.8. This toll on time can be decimated if a parallel processor is used. Here, if GPUs of larger number of cores are used, the processing delay is eliminated, and we get instantaneous results for the input. The number of cores required completely depends on the criticality of the application. GPUs such as Nvidia GeForce GTX 1050, AMD Radeon RX 570, or Nvidia GeForce GTX 1050 Ti can be used to meet the requirements.

## REFERENCES


[1] D.L. Donoho, Compressed sensing, IEEE Trans. Inform. Theory 52 (4) (2006) 1289–1306.

[2] S. Qaisar, et al., Compressive sensing: from theory to applications, a survey, J. Commun. Netw. 15 (5) (2013). 443456.

[3] L. Gan, Block compressed sensing of natural images, in: Proceedings of the International Conference on Digital Signal Processing (2007) 403–406.

[4] A.J. Jerri, The Shannon sampling theorem: its various extensions and applications: a tutorial review, Proc. IEEE 65 (11) (1977) 1565–1596.

[5] S. Foucart, H. Rauhut, A Mathematical Introduction to Compressive Sensing, vol. 1, Basel, Birkhuser, 2013.

[6] S. Mallat, A Wavelet Tour of Signal Processing: The Sparse Way, Academic Press, 2008.

[7] G. Kutyniok, Theory and aapplications of compressed sensing, GAMM-Mitteilungen 36 (1) (2013) 79–101.

[8] M. Aharon, M. Elad, A.M. Bruckstein, The K-SVD: an algorithm for designing of over-complete dictionaries for sparse representation, IEEE Trans. Signal Proc. 54 (2006). 43114322.

[9] J. Ma, S. Yu, Sparsity in compressive sensing, Lead. Edge 36 (8) (2017) 646–652.

[10] E.J. Cands, Mathematics of sparsity (and a few other things), in: Proceedings of the International Congress of Mathematicians, Seoul, South Korea, vol. 123, 2014.

[11] R. Rubinstein, M. Zibulevsky, M. Elad, Double sparsity: learning sparse dictionaries for sparse signal approximation, IEEE Trans. Signal Process. 58 (3) (2010) 1553–1564.

[12] M.S. Krishnamoorthy, An NP-hard problem in bipartite graphs, ACM SIGACT News. 7 (1) (1975) 26–26.

[13] L. Mancera, J. Portilla, L0-norm-based sparse representation through alternate projections, in: Image Processing, 2006 IEEE International Conference on IEEE, 2006, 2089–2092, Atlanta, GA, USA.

[14] A.M. Bruckstein, D.L. Donoho, A. Elad, From sparse solutions of systems of equations to sparse modeling of signals and images, SIAM Rev. 51 (2009) 3481.

[15] K.-H. Thung, P. Raveendran, A survey of image quality measures, in: Technical Postgraduates (TECHPOS), 2009 International Conference for IEEE, Kuala Lumpur, Malaysia, 2009, pp. 1–4.

[16] Image Guality Metrics: PSNR vs. SSIM Alain Hor MOIVRE, Département dinformatique, Facult des sciences, Universit de Sherbrooke Sherbrooke (Qubec), Canada, J1K2R1 Djemel Ziou.

[17] Q. Huynh-Thu, M. Ghanbari, Scope of validity of PSNR in image/video quality assessment, Electron. Lett. 44 (13) (2008) 800–801.

[18] Z. Wang, A.C. Bovik, Mean squared error: love it or leave it? IEEE Signal Process. Mag. 26 (2009) 98–117.

[19] Z. Wang, A.C. Bovik, H.R. Sheikh, E.P. Simoncelli, Image quality assessment: from error visibility to structural similarity, IEEE Trans. Image Process. 13 (4) (2004) 600–612.

[20] G.-H. Chen, C.-L. Yang, S.-L. Xie, Gradientbased structural similarity for image quality assessment, in: Image Processing, 2006 IEEE International Conference on IEEE, Atlanta, GA, USA, 2006, pp. 2929–2932.

[21] MATLAB and Statistics Toolbox Release, The MathWorks, Inc., Natick, MA, 2012.

[22] S. Matuska, R. Hudec, M. Benco, The comparison of CPU time consumption for image processing algorithm in Matlab and OpenCV, in: ELEKTRO, 2012 IEEE, 2012.

[23] P. Vogt, et al., Mapping spatial patterns with morphological image processing, Landsc. Ecol. 22 (2) (2007) 171–177.

[24] L. Lam, S.-W. Lee, C.Y. Suen, Thinning methodologies-a comprehensive survey, IEEE Trans. Pattern Anal. Mach. Intell. 14 (9) (1992) 869–885.

[25] J.H. Mathews, K.D. Fink, Numerical Methods Using MATLAB, vol. 4, Pearson Prentice Hall, Upper Saddle River, NJ, 2004.

[26] S. Makridakis, Accuracy measures: theoretical and practical concerns, Int. J. Forecast. 9 (4) (1993) 527–529.

CHAPTER

# REVIEW OF INTELLECTUAL VIDEO SURVEILLANCE THROUGH INTERNET OF THINGS

10

**Preethi Sambandam Raju[1,2], Murugan Mahalingam[1] and Revathi Arumugam Rajendran[3]**

[1]*Department of Electronics and Communication Engineering, Valliammai Engineering College, Chennai, India*
[2]*Department of Information and Communication Engineering, Anna University, Chennai, India* [3]*Department of Information Technology, Valliammai Engineering College, Chennai, India*

## CHAPTER OUTLINE



## 10.1 INTRODUCTION

Wireless multimedia sensor networks (WMSNs), outlet of networks, deal with audio–visual data and provide opulent information compared with traditional audio and video sensors through wireless sensor networks (WSNs) [1]. A subdivision of WMSNs, which has complete handling with video for surveillance purpose, is wireless video surveillance networks [2] that can also be a part of wireless video sensor networks when perceived from the implementation angle [3].

Surveillance can be achieved by monitoring the environment with the help of audio, sensing parameters as temperature, humidity, soil moisture, and sunlight, and through recording and analysis of video. Surveillance is now targeted for the delivery of high performance at low cost and low power consumption [4]. Video surveillance over Internet of Things (VS-IoT), a subdivision of multimedia Internet of Things (IoT), is an emerging category of the IoT that integrates image processing, deep learning, computer vision, artificial intelligence, and networks capability for use in motion detection, face recognition, behavior analysis, anomaly detection, event recognition, and

**The Cognitive Approach in Cloud Computing and Internet of Things Technologies for Surveillance Tracking Systems.**
**DOI: https://doi.org/10.1016/B978-0-12-816385-6.00010-6**

surveillance [5]. To the best of the authors' knowledge, this chapter is the first of its kind on the subject of the survey of video surveillance in terms of IoT networks.

The remainder of this chapter is arranged as follows. Section 2 deals with the classification of the IoT environment. Section 3 provides an introduction to the stages of VS-IoT along with a description of the sensing and monitoring techniques and analytics of the IoT, a discussion of data, and communication standards, and presents the different types of warehousing and an application design for VS-IoT. Section IV concludes this review chapter.

### 10.1.1 INTERNET OF THINGS ENVIRONMENTAL TAXONOMY

Pivoted on the IoT environmental features, taxonomy is comprised of three major building blocks, namely design, implementation, and processing. The three building blocks can further be subdivided into few specific feature blocks such as deployment types, design of architecture, sensor types, sensing time, tracking objects, and data requirements. The design block can be divided on the basis of deployment methods and architecture sorts [6–8]. Deployment methods can be centralized or in distributed fashion. Architecture for the IoT can be designed in two sorts as monolithic architecture and microservice architecture. The implementation block can be categorized on the basis of nature of sensing time and sensing types [9–11]. With an emphasis on sensing time, the sensor can be of either linear or nonlinear nature. Sensing in the IoT can be of either homogenous or heterogeneous types. The processing block can be classified depending on application types and data requirement modes [10,12]. An application of the IoT according to the tracking of objects can be of two types, namely: (1) single object tracking; and (2) multitarget tracking. Considering the data requirement modes, the environment can be categorized into a real-time mode and an offline mode. The taxonomy based on the IoT environment is summarized and depicted in Fig. 10.1.

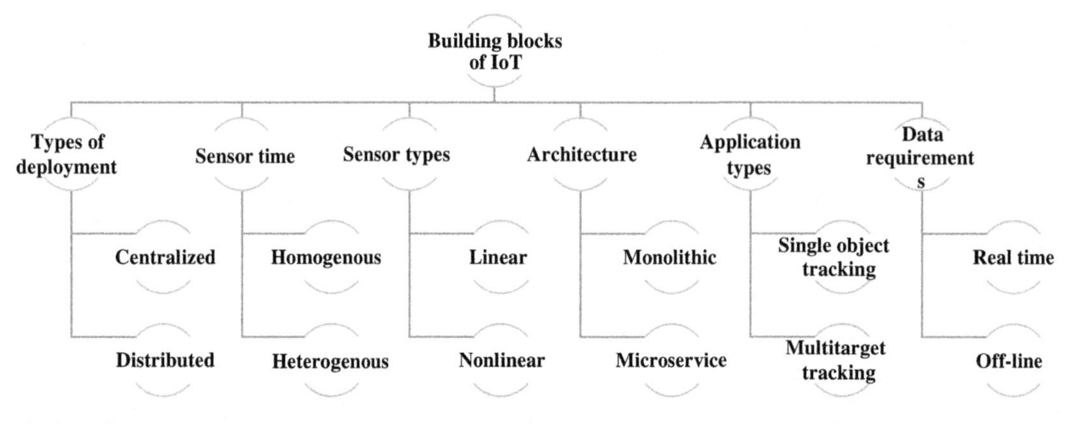


**FIGURE 10.1**

Internet of Things environmental taxonomy.

## 10.2 VIDEO SURVEILLANCE—INTERNET OF THINGS

The enactment of the VS-IoT networks comprises of five stages. The initial stage involves sensing and monitoring the environment, which is the soul of the VS-IoT structure. IoT data analytics, the second stage, handles the data generated by the preliminary stage. Following this is the communicating stage. In the third stage, the busted-out multimedia data are pushed to the data collection unit. The fourth stage indicates the five varieties of data warehousing for storage and computing. The concluding stage is an application-oriented design for the VS-IoT. This stage covers the techniques involved in the three main purposes of video surveillance. In Fig. 10.2, the survey of the VS-IoT along with the reference number of the chapter, the name of the author, and the year in which the technique has been proposed are portrayed in detail.

### 10.2.1 SENSING AND MONITORING

A portrayal of the VS-IoT indicates the first to form the VS-IoT environment that is the deployment of sensors and algorithms for motion detection. Based on employment variations, sensors can be of three major kinds, namely environmental layable, wearable, and implantable sensors [13]. Environmental layable sensors can be integrated with surroundings for applications like temperature and atmospheric pressure monitoring. Wearable sensors can be worn on the human body for pulse rate and other health parameters. Implantable sensors can be implemented in various devices and human organs for specific applications like vision sensor implantation on human eye retina. Some sensors are used discretely, and at times, they are clubbed to form a collaborative sensing platform for the VS-IoT [4]. In reality, it is found that motion detection befalls in substantial interludes. Thus it is obligatory to implement procedures for the perception of the motion. Motion detection can be achieved through sensors and algorithms. The combination of both algorithm and sensors efficiently gives rise to intelligent front-end devices. The challenge lies in implementing sensing and monitoring stages with low latency, low power consumption, and miniscule detection.

#### *10.2.1.1 Sensor-based motion detection*

Sensor implementation facilitates the work of motion detection. Based on the discrete or collaborative style of operation, a sensing platform can be of discrete or collaborative type. The sensors that can be worn on the human body are categorized as wearable body sensors (WBSs).

##### 10.2.1.1.1 Discrete sensing platform

Different types of sensors are used in video surveillance. Investigation of passive infra red sensors, rotational-directional sensors, and audio sensors form the subject matter of this chapter. The general PIR-based video surveillance mounts the PIR sensor on-board with a camera that turns on only when motion is detected by the PIR sensor, thus reducing the power consumed by the camera during idle times. Jeličić et al. [3] proposed a network that has a dense deployment of PIR sensors. These sensors send information about motion detection to the main node, which in turn switches on the specific camera based on PIR sensor's information and camera energy. This prototype achieves more power consumption compared with the previous one.

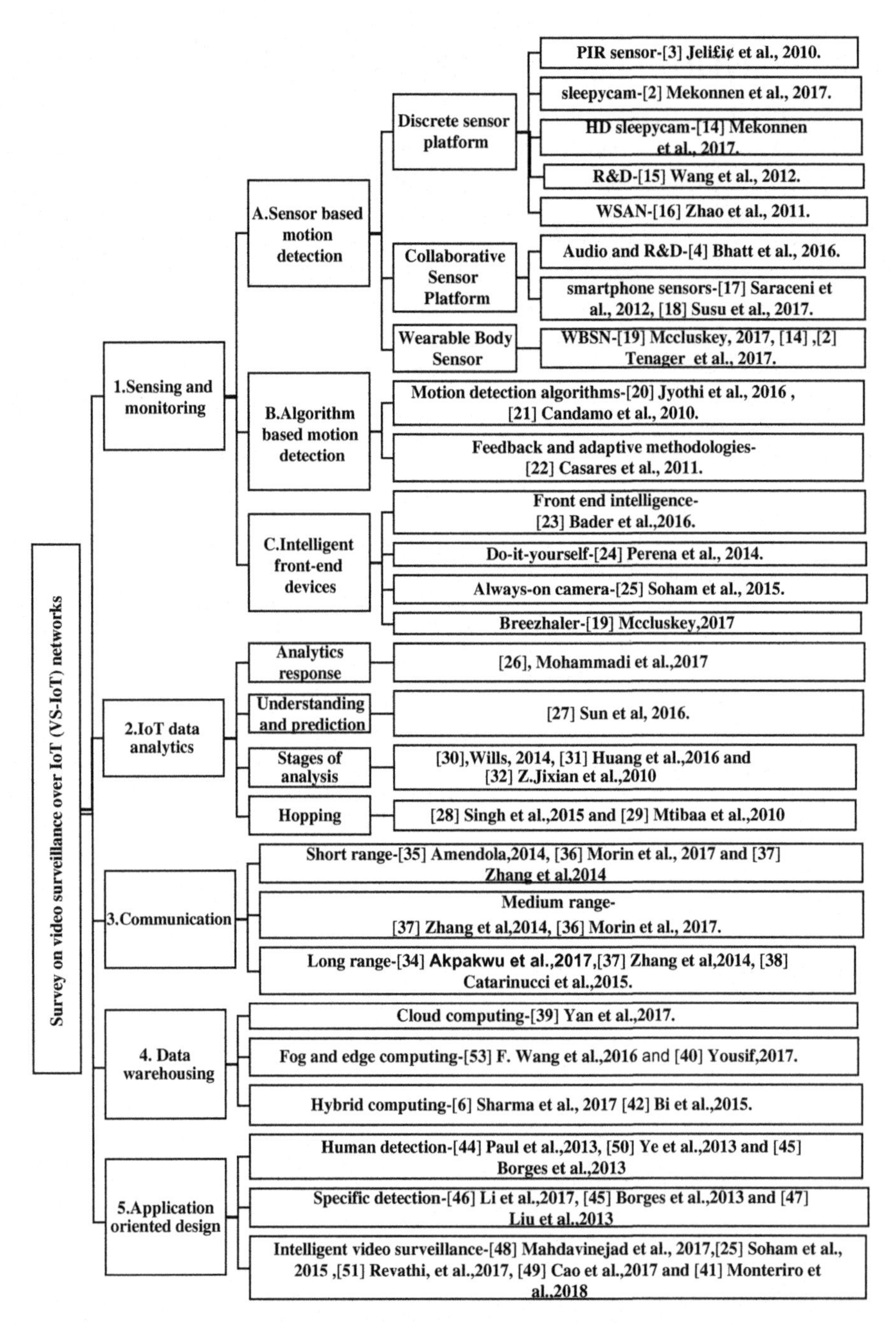


**FIGURE 10.2**

Survey on the video surveillance over Internet of Things.

Mekonnen et al. [2] proposed a sleepyCAM power management mechanism that involves the use of PIR sensor for motion detection and powering up of Raspberry Pi (RPi) using a relay. In sleepyCAM, normally RPi is powered off, and a step for proper shutdown of RPi achieves low power consumption during the waiting time of the surveillance application. As the earlier work [2], does not support high definition videos, Mekonnen et al. implemented the sleepyCAM power management mechanism for high definition videos using the Libelium Waspmote sensor platform. The monsoon power monitor tool is used for the measurement of the power consumption of both RPi and Waspmote [14]. Rotational and directional sensors have the capability to move the linked entity in a particular direction and to revolve for specific degrees. Thus connecting rotational and directional sensors to the video node helps position the camera and record the Event of absence. This reduces the power consumption of recording the unnecessary events and increases the camera coverage area [15]. A dense deployed audio sensor network, which is a low cost and low power device, is implemented to perform first level of noise detection, which in turn helps in motion detection. This wireless sensor audio network [16] overcomes the limitation of line of sight and lighting conditions in monitoring the environment.

##### 10.2.1.1.2 Collaborative sensing platform

The collaborative sensing platform uses the sensing capability of two or more sensors for the detection of the required actions. In Ref. [4], audio sensors and rotational and directional sensors are used together to record videos. At the initial stage, a dense deployed audio sensor is used for preliminary detection and, then on, audio detection sensors trigger the sparsely implemented video nodes to capture the videos. The audio sensor sends the location of motion to the rotational and directional sensor. Based on the location information, the rotational and directional sensor revolves and moves the video node to a particular direction and angle to capture the videos. Currently, the surveillance system is explored with the help of sensors in smartphones to enable low cost and ease implementation [17,18].

##### 10.2.1.1.3 Wearable body sensors

Wireless sensors can be wearable, giving rise to a WBS that has invented wireless ears and wireless tongues. It is not that all body sensors must be wearable. They can also be linked to a function along with some organs of human body. One such example is the sound signature technology-based Breezhaler by Novartis, which works on the basis of inhaling and exhaling of air through nose [19]. Despite many advantages of WBSs, battery life is the main drawback.

Sleepy devices introduced into the system help in overcoming this weakness [3,14]. In sleepy devices, the wake-up perception is used as an alternative for providing constant power. In wake-up concept, power is provided to the right required devices of the system at its right time and then is laid into either sleep or shutdown mode when the devices are redundant. WBSs also suffer from challenges such as the dimensions of WBSs and health disputes due to enactment on the human body.

#### 10.2.1.2 Algorithm-based motion detection

Motion detection algorithm determines the movement of an object through a comparison of two or more successive images in the video. In the absence of any motion, the video is neither stored nor processed. Otherwise, the current frame is processed in the event of motion detection [20]. Motion detection traditionally involves background subtraction, temporal difference, optical flow, and hybrid methods [21]. Generally, motion recognition is meant to execute foreground object detection and trace each frame independently. However, in the feedback method [22], the data from tracking stage are integrated into the feedback stage, and thus foreground detection is performed in smaller regions compared with the whole frame. Thus it results in dropping the handling energy of a frame and switches on to an idle state at the end of processing a frame without causing tracking failure. An adaptive method [22] is adopted to drop frames even when there is no empty scene. The idle state duration is adaptively changed based on the amount of activity in the scene and tracking the object speed.

#### 10.2.1.3 Intelligent front-end devices

Intelligence requires coming with front-edged devices for a complete enjoyment of the IoT and making them smart IoT front-end devices [23]. This idea of smart IoT front-end devices cuts down unnecessary data movement. Front-end devices can be made of software-driven architecture for tuning the parameters to get the desirable output. This notion is known as Do-It-Yourself [24]. A smart front-end camera with a decision-making algorithm makes judicious decisions when to store or transmit a video [25]. IBM attempts at the creation of a smart front-end IoT device named Cognitive Hypervisor. This device extracts noteworthy patterns from numerous data collected by various WBSs for enhanced doctor–patient interaction [19]. There is a need for invention and research of innovative sensors.

### 10.2.2 INTERNET OF THINGS DATA ANALYTICS

The technique of using the data sets according to the necessities is termed as data analytics. IoT data can be classified on the basis of aspects such as analytics response, understanding and prediction of analytics, analysis stages, and the number of hops, as portrayed in Fig. 10.3. IoT data can be fast data and big data based on the quickness of analytics response [26]. Fast data represent analytics that should be done in the source of data within milliseconds for fast and judicious decisions. The IoT devices yield countless data that upshot to big data. These big data must first be extracted

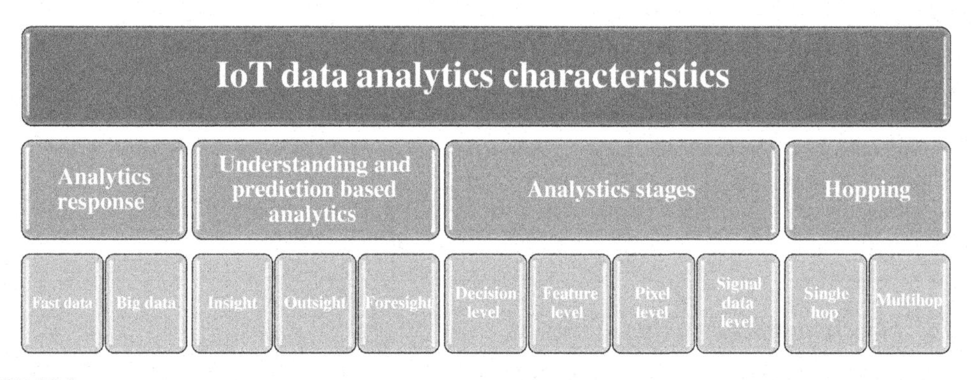


**FIGURE 10.3**

Internet of Things data analytics classification based on characteristics.

from the devices. Insight is drawn from the collected data, and then the analytics is performed. These big data can be categorized into 6Vs (velocity, variety, volume, value, veracity, and variability). IoT data analytics based on understanding and prediction can be carried out on three levels, namely insight, outsight, and foresight [27]. Insight analytics is executed for an in-depth understanding of its own data, while outsight analytics is for understanding the external aspects. Foresight analytics is implemented for understanding, prediction, and prevention.

Data analytics takes place in four stages in the IoT embedding, namely decision level, feature level, pixel level, and signal level [10]. Decision level evaluates the collected data for a judicious conclusion. Feature level analyzes the miniscule data for feature classification. Pixel level details are examined in pixel level analytics. Signal level studies signal data values. Data analytics based on hopping perspective can be of two types, namely single hop and multihop [28,29]. In single hop, data are directly sent to the data collection center, whereas in multihop, data are passed through the adjacent centers to reach the data center. Table 10.1 encapsulates the classification based on the aspects of data analytics. The biggest challenge lies in the fusion of massive heterogeneous data and intensive intelligent computing.

### 10.2.3 COMMUNICATION

There are numerous traditional techniques for data communication. Based on the coverage range, communication can be of three types, namely short-range communication [33], medium-range communication, and long-range communication [34]. In the short-range communication, technologies

**Table 10.1 Classification Based on Data Analytics Aspect.**

| Author and Year of Publishing | Characteristics of Data Analytics | Types/Stages | Examples |
|---|---|---|---|
| Mohammadi et al. (2017) [26] | Analytics response | Fast data | Emergency situation with streaming data analytics |
| | | Big data | Smart city monitoring |
| Sun et al. (2016)[27] | Understanding and prediction | Insight | Current scenario |
| | | Outsight | Weather impact |
| | | Foresight | Expected natural calamities |
| Wills (2014) [30], Huang et al. (2016) [31], and Zhang (2010) [32] | Analytics stages | Decision level | Healthcare |
| | | Feature level | Activity recognition, fall detection, and sentiment analysis |
| | | Pixel level | Sign language data set and feature recognition |
| | | Signal data level | Controlling and monitoring the environment, and human–computer interfaces |
| Singh et al. (2015) [28] and Mtibaa et al. (2017) [29] | Hopping | Single hop | Centralized environment |
| | | Multihop | Distributed environment |

**Table 10.2 VS-IoT Communication Classification.**

| Author and Year of Publishing | Range | Techniques | | Challenges for Implementing in IoT Scenario |
|---|---|---|---|---|
| Amendola (2014) [35], Morin et al. (2017) [36], and Zhang et al. (2014) [37] | Short range | Radio-frequency identification | Passive tags | Increase reader coverage area |
| | | | Active tags | Development for large-scale applications |
| | | IEEE 802.15.1 | Classic Bluetooth | High duty cycle<br>High transmission power, security, and small distance |
| | | | BT-LE | Small data transmission |
| | | IEEE 802.15.4 ZigBee | | Low bandwidth of 250 kbps |
| | | IEEE 802.15.6 BSN | | Poor computation capability |
| Zhang et al. (2014) [37] and Morin et al. (2017) [36] | Medium range | IEEE 802.11 WLAN | | Mobility<br>Coverage and energy efficiency |
| | | IEEE 802.16 WiMax | | High cost<br>Poor bandwidth due to large clients. Weather conditions have a big impact |
| | | IEEE 802.20 MBWA | | Does not support sensor networks, thus put to slumber |
| Akpakwu et al. (2017) [34], Zhang et al. (2014) [37], and Catarinucci et al. (2015) [38] | Long range | IEEE 802.22 Cognitive radio | | This possibility is difficult for WSN applications |
| | | SIGFOX | | Low data rate |
| | | LoRa | | Lower lifetime compared with many low power technologies |
| | | NB-IoT | | To encounter the varied necessities of the IoT |
| | | Hybrid techniques | | Coordinated research activities in both RFID and WSN challenge areas |

*BSN*, Body sensor networks; *LoRa*, long range; *NB-IoT*, narrowband Internet of Things; *VS-IoT*, video surveillance over Internet of Things.

with a coverage of up to 100 m are considered, whereas a coverage of about 10 km is grouped in the medium-range communication. The exposure range is about 100 km in the long-range communication. The communication classification for the VS-IoT is detailed in Table 10.2.

#### 10.2.3.1 Short-range communication

Short-range communication includes technologies such as radio-frequency identification (RFID), Bluetooth, ZigBee, and body sensor networks (BSNs) [35–37]. RFID is a low cost and low power technology that can work with or without a battery. This system is compatible with abundant applications. RFID has tags that have the ability to transmit information, consuming less or no power and thus facilitating extension of their lifetime to decades. The RFID tags can be either passive tags or battery-assisted active tags. RFID technology is easy to implement and maintain. In spite of the zero power tags, the disadvantage of RFID come from the point that it can operate only in the reader coverage area, that is, up to 5 m if fully passive tags are used; else up to 25 m if battery-assisted tags are used. Feasibility of RFID in some applications has been a matter of doubt due to the emission of electric field and the power absorption of the human body from the tags [35]. The key challenges in implementing RFID increase the reader coverage area and limit the emission of electric field from RFID devices. The IEEE 802.15.1 standard comprises of classic Bluetooth and Bluetooth Low Energy (BT-LE). Classic Bluetooth is of low cost and easy to install, whereas BT-LE consumes low power for operation and can operate in two modes. IEEE 802.15.4 ZigBee is useful for the creation of a self-configurable and self-healing network. IEEE 502.15.6 BSN is beneficial to form a network with wearable sensors. A BSN delivers mobility and energy efficient network.

#### 10.2.3.2 Medium-range communication

Medium-range communication takes account of techniques, for instance, wireless local area networks (WLANs), worldwide interoperability for microwave access (WiMAX), and mobile broadband wireless access (MBWA) [36,37]. IEEE 802.11 WLAN provides a continuous high throughput connection. IEEE 802.16 WiMAX provides a secure mobile network with high transmission speed. IEEE 802.20 MBWA offers high velocity and challenges to the upcoming 3G technologies.

#### 10.2.3.3 Long-range communication

Long-range communication consists of methods like cognitive radio, SIGFOX, long range (LoRa), and narrowband Internet of Things (NB-IoT). IEEE 802.22 cognitive radio has an option to use licensed bands by unlicensed users on a noninterference basis. SIGFOX is easy to deploy and connect well to underground buried devices. LoRa supports ultralow traffic intensity. NB-IoT provides improved bidirectional information compared with most of the unlicensed low power wireless access. Although WSNs, when compared with RFID, provide a low cost ad-hoc network to analyze and control the environment, most of the WSN motes consume substantial power that shrinks the lifetime of the network. Therefore hybrid work combining RFID and WSN is evolved, to enable the use of only RFID in the usual conditions and for the timely use of WSN in the circumstances of emergency for informing the required individuals and henceforth able to manage the power efficiently [38].

### 10.2.4 DATA WAREHOUSING

The data acquired through the sensors should be warehoused for workout and other indispensable operations. Privacy and security are closely associated with each other, and the collected data should be secured from hackers. By means of illustration in Table 10.3, the warehousing of data can be done in the succeeding five techniques.

**Table 10.3 Overview of Five Types of Data Warehousing.**

| Author and Year of Publishing | Hetero/Homo Technology | Classification | | Challenges for Research |
|---|---|---|---|---|
| Yan et al. (2017) [39], Yousif (2017) [40], and Jianhua et al. (2018) [41] | Types based on homotechnology | Cloud | | Data privacy<br>Verification on correctness of data processing<br>Limited computations |
| | | Fog | Ad-hoc<br>Dedicated | Lack of concrete methods, tools, and frameworks |
| | | Edge | | Is not efficient for static<br>Limited data processing |
| Sharma et al. (2017) [6] and Bi et al. (2015) [42] | Types based on hybrid technology | Collaborative computing | | Interplay with different types of computing must be developed at high rates |
| | | Crowd computing | | Mobility of devices for data sharing<br>Temporal and spatial correlations |

#### 10.2.4.1 Cloud

Cloud is wireless centered computing that provides the provision of rearranging, processing, storing the data sets, and providing the required content to the users based on their demands [39]. Cloud computing provides scalable, elastic, fault-tolerant, and pay-per-use architecture. As it is a centralized architecture, it may lead to a situation like data loss during large data transmissions from the devices to the cloud. Hence, there is a need to keep the data close to the sensor devices.

#### 10.2.4.2 Fog and edge

The concept of computing data close to sensor devices leads to two types of computing, namely fog computing and edge computing [40]. Fog computing is meant to bring the small part of the cloud close to the IoT gateways. This leads to a distributed architecture and overcomes the partial disadvantages of the cloud. The main benefits of fog computing are heterogeneity, interplay with cloud computing, high mobility, and latency reduction. Fog can be either ad-hoc or dedicated fogs based on their computing resources. The legal requirement to have the data near the devices, latency between edge devices and gateways, results in ineffective decisions that, in turn, lead to edge computing. Edge computing does the storing and processing in edge devices' bottom of the architecture hierarchy. It also offers location-aware computing for real-time analytics with reduced network traffic.

#### 10.2.4.3 Hybrid technologies

A combination of the above computing types has been the melting pot for the evolution of two types of computing, namely collaborative computing and crowd computing [6]. Collaborative computing refers to performing analytics by combining two or three of homocomputing technologies. It provides a feedback to the end users and combines the advantages of homocomputing technologies. Crowd computing uses human intelligence (crowd) along with one, two, or three of the above computing techniques for the delivery of higher computing capability. In this type, nearby devices come into a social tie and share their resources.

### 10.2.5 APPLICATION-ORIENTED DESIGN

Surveillance is given intelligence and is designed to work for a specific purpose. Video surveillance is mostly done for human detection [43]. Surveillance is also done for some specific detection and for intelligent analyses. Generally, human detection in videos is done by detection followed by a classification [44]. The object detection can be of two different types based on static and dynamic backgrounds. Object classification done following object detection is divided into three types,

**Table 10.4 Summary of Application-Oriented Design.**

| Author and Year of Publishing | Video Surveillance Purpose | Types and Subtypes | | Supporting Technologies |
|---|---|---|---|---|
| Paul et al. (2013) [44], Ye et al. (2013) [50], and Borges et al. (2013) [45] | Human detection | Object detection | Static background | Background subtraction based on a Gaussian mixture, model region segmentation-based graph cut, and compressive sensing |
| | | | Dynamic background | Lucas–Kanade–Tomasi tracker, mean shift, and level set contour |
| | | Object classification | Shape-based | Standard template matching |
| | | | Motion-based | Self-similarity-based time–frequency technology and optical flow based |
| | | | Texture-based | Histograms of oriented gradient and SVM |
| Li et al. (2017) [46], Borges et al. (2013) [45], and Liu et al. (2013) [47] | Specific detection | Activity recognition | | SVM, *k*-nearest neighbor, and Bayes classifier |
| | | Silhouettes extraction | | Background segmentation approaches and traditional classifiers |
| | | 3D sensing | | Intelligent fiber-grating-based 3D vision sensory system |
| Mahdavinejad et al. (2017) [48], Soham et al. (2015) [25], Revathi et al. (2017) [51], Cao et al. (2017) [49] | Intelligent video surveillance | Machine learning algorithms | | Smart Monitoring and anomaly detection |
| | | Deep learning algorithms | | Restricted Boltzmann machine algorithm and convolution neural networks |
| | | Classifiers | | SVM, Naïve–Bayes, and decision tree |

*SVM*, Support vector machine.

namely shape-based, motion-based, and texture-based. Video surveillance can also be used for specific purposes like action recognition, silhouette extraction, and 3D sensing [45–47]. Machine learning algorithms [48], deep learning algorithm [25], and classifiers [49] are used in the process of adding intelligence to the video surveillance system. A detailed taxonomy for the video surveillance purpose is depicted in Table 10.4. The challenge lies in the convergence of suitable deep learning techniques, artificial intelligence, and things to attain the targeted application. The IoT is in the early stage of implementation and faces lot of challenges itself [52]. An analysis of the overview of the VS-IoT leads to the inference of its notable three challenges, in the areas of energy depletion, informal latency, and multimodal data quality. In general, a prototype developed for processing high quality multimedia data results in high power consumption, and the steps that are taken to reduce the power consumption may result in a low standard of video and can add on latency. Thus, in developing a VS-IoT, it is crucial to keep confrontation on both energy consumption and video quality. Information latency also plays a major role, as delayed data become useless data in times of emergencies. Hence, the VS-IoT leads to research options in implementing a low energy, high quality, and reduced latency system.

## 10.3 CONCLUSION

This chapter has taken up the subject of the elucidation of IoT environmental taxonomy on the aspect of design, implementation, and processing. The survey for VS-IoT networks has been investigated stage by stage. VS-IoT has been interpreted as comprising five stages, namely sensing and monitoring, IoT data analytics, communication, data warehousing, and application-oriented design. The survey was done starting from the initial stage leading to application-oriented design with the objective of a rich knowledge relating to the implementation of VS-IoT and obtaining a mature framework for the same. Sensing and monitoring stage is analyzed on the basis of sensor usage, algorithm, and intelligent front-end devices. IoT data analytics has been discussed and related with some key aspects. Illustration of the communication stage has been developed based on a coverage distance. Five types of data warehousing have been elaborated. Then, the application-oriented design is summarized based on human detection, specific detection, and intelligent video surveillance. An overview of the challenges faced by the video surveillance system when implemented through the IoT networks has been presented in addition to these seen during each stage.

## REFERENCES


[1] T. Mekonnen, P. Porambage, E. Harjula, M. Ylianttila, Energy consumption analysis of high quality multi-tier wireless multimedia sensor network, IEEE Access 5 (2017) 15848–15858.

[2] T. Mekonnen, E. Harjula, T. Koskela, M. Ylianttila, sleepyCAM: power management mechanism for wireless video-surveillance cameras, in: 2017 IEEE International Conference on Communications Workshops (ICC Workshops), Paris, 2017, pp. 91–96.

[3] V. Jeličić, M. Magno, D. Brunelli, V. Bilas, L. Benini, An energy efficient multimodal wireless video sensor network with eZ430–RF2500 modules, in: 5th International Conference on Pervasive Computing and Applications, Maribor, 2010, pp. 161–166.

[4] R. Bhatt, R. Datta, A two-tier strategy for priority based critical event surveillance with wireless multimedia sensors, Wirel. Netw. 22 (1) (2016) 267–284.
[5] C. Long, Y. Cao, T. Jiang, Q. Zhang, Edge computing framework for cooperative video processing in multimedia IoT systems, IEEE Trans. Multimedia 20 (5) (2018) 1126–1139. Available from: https://doi.org/10.1109/TMM.2017.2764330.
[6] S.K. Sharma, X. Wang, Live data analytics with collaborative edge and cloud processing in wireless IoT networks, IEEE Access 5 (2017) 4621–4635.
[7] T.W. Martin, K. Chang, A data fusion formulation for decentralized estimation predictions under communications uncertainty, in: 2006 9th International Conference on Information Fusion, Florence, 2006, pp. 1–7.
[8] L. Sun, Y. Li, R.A. Memon, An open IoT framework based on microservices architecture, China Commun. 14 (2) (2017) 154–162.
[9] J.W. Elmenreich, R. Leidenfrost, Fusion of heterogeneous sensors data, in: 2008 International Workshop on Intelligent Solutions in Embedded Systems, Regensburg, 2008, pp. 1–10.
[10] F. Alam, R. Mehmood, I. Katib, N.N. Albogami, A. Albeshri, Data fusion and IoT for smart ubiquitous environments: a survey, IEEE Access 5 (2017) 9533–9554.
[11] U. Rashid, H.D. Tuan, P. Apkarian, H.H. Kha, Multisensor data fusion in nonlinear Bayesian filtering, in: 2012 Fouth International Conference of Communication Electronics, 2012, pp. 351–354.
[12] M. Marjani, F. Nasaruddin, A. Gani, A. Karim, I.A.T. Hashem, A. Siddiqa, et al., Big IoT data analytics: architecture, opportunities, and open research challenges, IEEE Access 5 (2017) 5247–5261.
[13] D. He, S. Zeadally, An analysis of RFID authentication schemes for Internet of Things in healthcare environment using elliptic curve cryptography, IEEE Internet Things J. 2 (1) (2015) 72–83.
[14] T. Mekonnen, E. Harjula, A. Heikkinen, T. Koskela, M. Ylianttila, Energy efficient event driven video streaming surveillance using sleepyCAM, in: Proceedings of the 2017 IEEE International Conference in Computer and Information Technology, 2017, pp. 107–113.
[15] Y.C. Wang, Y.F. Chen, Y.C. Tseng, Using rotatable and directional (R&D) sensors to achieve temporal coverage of objects and its surveillance application, IEEE Trans. Mobile Comput. 11 (8) (2012) 1358–1371.
[16] G. Zhao, H. Ma, Y. Sun, H. Luo, X. Mao, Enhanced surveillance platform with low-power wireless audio sensor networks, in: 2011 IEEE International Symposium on a World of Wireless, Mobile and Multimedia Networks, Lucca, 2011, pp. 1–9.
[17] S. Saraceni, A. Claudi, A.F. Dragoni, An active monitoring system for real-time face-tracking based on mobile sensors, in: Proceedings ELMAR-2012, Zadar, 2012, pp. 53–56.
[18] A.E. Susu, Low-cost distributed video surveillance with discarded mobile phones, in: 2017 21st International Conference on Control Systems and Computer Science (CSCS), Bucharest, 2017, pp. 279–286.
[19] B. Mccluskey, Wear it well, Eng. Technol. 12 (1) (2017) 32–35.
[20] S.N. Jyothi, K.V. Vardhan, Design and implementation of real time security surveillance system using IoT, in: 2016 International Conference on Communication and Electronics Systems (ICCES), Coimbatore, 2016, pp.1–5.
[21] J. Candamo, M. Shreve, D.B. Goldgof, D.B. Sapper, R. Kasturi, Understanding transit scenes: a survey on human behavior-recognition algorithms, IEEE Trans. Intell. Transp. Syst. 11 (1) (2010) 206–224.
[22] M. Casares, S. Velipasalar, Adaptive methodologies for energy-efficient object detection and tracking with battery-powered embedded smart cameras, IEEE Trans. Circuits Syst. Video Technol. 21 (10) (2011) 1438–1452.
[23] A. Bader, H. Ghazzai, A. Kadri, M.S. Alouini, Front-end intelligence for large-scale application-oriented Internet-of-Things, IEEE Access 4 (2016) 3257–3272.

[24] C. Perera, C.H. Liu, S. Jayawardena, M. Chen, A survey on Internet of Things from industrial market perspective, IEEE Access 2 (2014) 1660–1679.
[25] S.J. Desai, M. Shoaib, A. Raychowdhury, An ultra-low power, "Always-On" camera front-end for posture detection in body worn cameras using restricted boltzman machines, IEEE Trans. Multi-Scale Comput. Syst. 1 (4) (2015) 187–194.
[26] M. Mohammadi, A. Al-Fuqaha, S. Sorour, M. Guizani, Deep learning for IoT big data and streaming analysis: a survey, IEEE COM S&T J. (2017) 1–34.
[27] Y. Sun, H. Song, A.J. Jara, R. Bie, Internet of Things and big data analytics for smart and connected communities, IEEE Access 4 (2016) 766–773.
[28] A.K. Singh, S. Rajoriya, S. Nikhil, T.K. Jain, Design constraint in single-hop and multi-hop wireless sensor network using different network model architecture, in: 2nd International Conference on Computing, Communication & Automation, Noida, 2015, pp. 436–441.
[29] A. Mtibaa, A. Emam, K.A. Harras, On practical multihop wireless communication: insights, limitations, and solutions, in: 2017 IEEE 13th International Conference on Wireless and Mobile Computing, Networking and Communications (WiMob), Rome, 2017, pp. 1–8.
[30] M.J. Wills, Decision through data: analytics in healthcare, J. Healthc. Manag. 57 (4) (2014) 254–262.
[31] C.W. Huang, S. Narayanan, Comparison of feature-level and kernel-level data fusion methods in multisensory fall detection, in: 2016 IEEE 18th International Workshop on MultimediaSignal Processing (MMSP), Montreal, QC, 2016, pp. 1–6.
[32] Z. Jixian, Multi-source remote sensing data fusion: status and trends, Int. J. Image Data Fusion 1 (1) (2010) 5–24.
[33] B.N. Silva, M. Khan, K. Han, Internet of Things: a comprehensive review of enabling technologies, architecture, and challenges, IETE Tech. Rev. (2017). Available from: https://doi.org/10.1080/02564602.2016.1276416.
[34] G.A. Akpakwu, B.J. Silva, G.P. Hancke, A.M. Abu-Mahfouz, A survey on 5G networks for the Internet of Things: communication technologies and challenges, IEEE Access 6 (2018) 3619–3647.
[35] S. Amendola, R. Lodato, S. Manzari, C. Occhiuzzi, G. Marrocco, RFID technology for IoT-based personal healthcare in smart spaces, IEEE Internet Things J. 1 (2) (2014) 144–152.
[36] É. Morin, M. Maman, R. Guizzetti, A. Duda, Comparison of the device lifetime in wireless networks for the Internet of Things, IEEE Access 5 (2017) 7097–7114.
[37] Y. Zhang, L. Sun, H. Song, X. Cao, Ubiquitous WSN for healthcare: recent advances and future prospects, IEEE Internet Things J. 1 (4) (2014) 311–318.
[38] L. Catarinucci, D. de Donno, L. Mainetti, L. Palano, L. Patrono, M.L. Stefanizzi, et al., An IoT-aware architecture for smart healthcare systems, IEEE Internet Things J. 2 (6) (2015) 515–526.
[39] Z. Yan, X. Yu, W. Ding, Context-aware verifiable cloud computing, IEEE Access 5 (2017) 2211–2227.
[40] M. Yousif, Cloudy, foggy and edgy, IEEE Cloud Comput. 4 (2) (2017) 4–5.
[41] A. Monteiro, M. de Oliveira, R. de Oliveira, T. da Silva, Embedded application of convolutional neural networks on Raspberry Pi for SHM, Electron. Lett. 54 (11) (2018) 680–682. Available from: https://doi.org/10.1049/el.2018.0877.
[42] S. Bi, R. Zhang, Z. Ding, S. Cui, Wireless communications in the era of big data, IEEE Commun. Mag. 53 (10) (2015) 190–199.
[43] A.R. Revathi, D. Kumar, A survey of activity recognition and understanding the behavior in video surveillance, Adv. Comput. Sci. Inf. Technol. (2012) 375–384.
[44] M. Paul, S.M. Haque, S. Chakraborty, Human detection in surveillance videos and its applications—a review, EURASIP J. Adv. Sig. Proc. 176 (1) (2013) 1–16.
[45] P.V.K. Borges, N. Conci, A. Cavallaro, Video-based human behaviour understanding: a survey, IEEE Trans. Circuits Syst. Video Technol. 23 (11) (2013) 1993–2008.

[46] Y. Li, R. Xia, Q. Huang, W. Xie, X. Li, Survey of spatio-temporal interest point detection algorithms in video, IEEE Access 5 (2017) 10323–10331.

[47] H. Liu, S. Chen, N. Kubota, Intelligent video systems and analytics: a survey, IEEE Trans. Ind. Inform. 9 (3) (2013) 1222–1233.

[48] M.S. Mahdavinejad, M. Rezvan, M. Barekatain, P. Adibi, P. Barnaghi, A.P. Sheth, Machine learning for Internet of Things data analysis: a survey, Digital Commun. Networks 4 (3) (2018) 161–175.

[49] N. Cao, S.B. Nasir, S. Sen, A. Raychowdhury, Self-optimizing IoT wireless video sensor node with in-situ data analytics and context-driven energy-aware real-time adaptation, IEEE Trans. Circuits Systems I: Regular Papers 64 (9) (2017) 2470–2480.

[50] Y. Ye, S. Ci, A.K. Katsaggelos, Y. Liu, Y. Qian, Wireless video surveillance: a survey, IEEE Access 1 (2013) 646–660.

[51] A.R. Revathi, D. Kumar, An efficient system for anomaly detection using deep learning classifier, Signal Image Video P. 11 (2) (2017) 291–299.

[52] F. Wang, L. Hu, J. Hu, J. Zhou, K. Zhao, Recent advances in the Internet of Things: multiple perspectives, IETE Techn. Rev. 34 (2) (2016) 122–132.

CHAPTER

# VIOLENCE DETECTION IN AUTOMATED VIDEO SURVEILLANCE: RECENT TRENDS AND COMPARATIVE STUDIES

# 11

**S. Roshan, G. Srivathsan, K. Deepak and S. Chandrakala**

*Department of Computer Science, School of Computing, SASTRA Deemed to be University, Thanjavur, India*

## CHAPTER OUTLINE



## 11.1 INTRODUCTION

The rapid growth in the amount of video data has led to the increasing need for surveillance and anomaly detection. Such anomalous events rarely occur as compared with normal activities. Therefore to lessen the waste of labor and time, developing automated video surveillance systems for anomaly detection has become the need of the hour. Detection of abnormalities in videos is a challenging task as the definition of anomaly can be ambiguous and vaguely defined. They vary widely based on the circumstances and the situations in which they occur. For example, riding a bicycle in a regular pathway is a normal activity, but doing the same in a walk-only lane should be flagged as anomalous. The irregular internal occlusion is a notable yet challenging feature to

The Cognitive Approach in Cloud Computing and Internet of Things Technologies for Surveillance Tracking Systems.
DOI: https://doi.org/10.1016/B978-0-12-816385-6.00011-8

describe the anomalous behavior. In addition, representation of video data and its modeling induce more difficulty due to its high dimensionality, noise, and highly varying events and interactions. Other challenges include illumination variations, viewpoint changes, camera motions, and so on.

One of the significant aspects of anomaly detection includes violence recognition and detection. The increase in threats to security around the world makes the use of video cameras to monitor people necessary, and thereby early detection and recognition of these violent activities could greatly reduce these risks. The modeling techniques used for anomaly or violent detection can be broadly classified as shallow and deep models. The main objective of our paper is to perform a comparative study of the above-mentioned models.

Shallow modeling techniques are those that are not capable of learning features on their own but rather features extracted using handcrafted methods must be provided to a shallow network for their classification. A shallow network can be classifier models like support vector machine (SVM), artificial neural network (ANN) with one hidden layer, and so on. These models are best suited for supervised learning, which in the given data should be well labeled. The main drawback of this modeling technique is that they do not adapt to pattern changes automatically. Also the labeling process can be manually intensive. Lloyd et al. [1] have proposed a real-time descriptor that models crowd dynamics for anomaly detection by encoding changes in crowd texture using temporal summaries of gray-level cooccurrence matrix features, in which *k*-fold cross validation was performed for training a random forest classifier. Their proposed method outperforms the state-of-the-art results over the UMN, UCF, and Violent Flows (ViF) data sets. Similarly, Bilinski and Bremond [2] have used an extension of improved Fisher vectors (IFVs), which allows the videos to be represented using both local features and their spatio-temporal positions for violence recognition and detection. Their results have shown significant improvement in four publicly available standard benchmark data sets.

In contrast to shallow models, most of the deep models do not require a separate feature extractor, as they are based on the feature learning technique, which is that they learn their own features from the given data and classify based on them. In addition, apart from end-to-end learning, the above extracted features can be given as input to the SVMs and other shallow model classifiers. Another way to implement deep models is by using the features from handcrafted feature descriptors and providing it to a deep classifier. These models work on both supervised and unsupervised learning-based methods but are better suited for the latter. Even though they work with unlabeled data, they require high volumes of data and computational power. Chong and Tay [3] propose a convolutional spatio-temporal autoencoder to learn the regular patterns in the training videos for anomaly detection. Even though the model can detect abnormal events and is robust to noise, depending on how complex the activity is, more false alarms may occur. One other work on this model is proposed by Sudhakaran and Lanz [4], in which a convolutional long short-term memory (CLSTM) is used to train a model for violence detection. On comparing this method with other state-of-the-art techniques, their proposed method shows a promising result on the used data sets. A general system for abnormality or violent detection is shown in Fig. 11.1.

In this chapter, we wish to compare and analyze the above shallow and deep models based on their performance. The content of the chapter is as follows: Section 11.2 presents the recent and promising feature detectors used in anomaly and violence detection tasks, Section 11.3 discusses recent works in anomaly detections and a few methods that have been proven to be promising for our task, and the experimentation and analysis part of this chapter is dealt in Section 11.4.

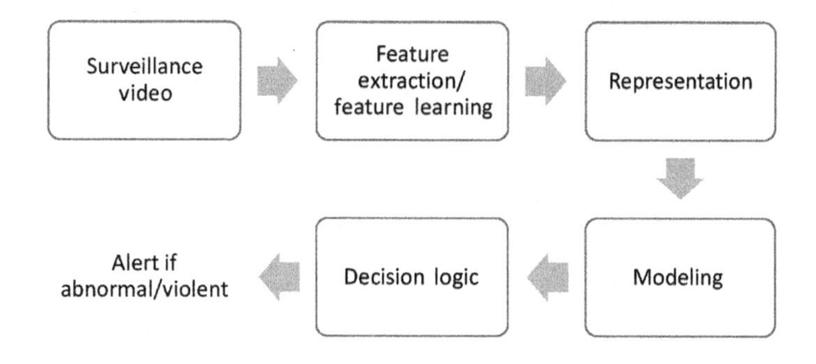


**FIGURE 11.1**

Overview of violence detection system.

## 11.2 FEATURE DESCRIPTORS

This section discusses about the feature descriptors used in our studies and also other recent state-of-the-art descriptors.

### 11.2.1 HISTOGRAM OF ORIENTED GRADIENTS

Histogram of oriented gradients (HOGs) is a feature descriptor for object detection and localization, which can compete with the performance given by deep neural networks. In HOG, the distribution of the directions of the gradients is used as features. This is due to the fact that edges and corners have high variations in intensities, and hence, calculating the gradient along with the directions can help detect this information from the image.

### 11.2.2 SPACE–TIME INTEREST POINTS

By extending the Harris detector, Laptev and Lindeberg [5] and Laptev [6] proposed the space–time interest point (STIP) detector. After extracting the points with large gradient magnitude with the help of a 3D Harris corner detector, a second-moment matrix is computed for each spatio-temporal interest points. The features obtained from this descriptor are used to characterize the spatio-temporal, local motion, and appearance information in volumes.

### 11.2.3 HISTOGRAM OF ORIENTED OPTICAL FLOW

Due to the relative motion between an observer (camera) and a scene (image, video), a pattern of apparent motion of objects, surfaces, and edges is created. This is called as *Optical Flow*. Histogram of oriented optical flow (HOF) [7] is a feature based on the optical flow that represents the sequence of actions at each instance of time. It is scale-invariant and independent of the direction of motion.

### 11.2.4 VIOLENCE FLOW DESCRIPTOR

One important feature descriptor is the violence flow, which uses the frequencies of quantized values in a vectorized form. This is different from other descriptors in a way that, rather than considering magnitudes of temporal information, the comparison of the magnitudes is taken for each as it gives much more meaningful measures in terms of the predecessor frame [8]. Instead of using local appearances, the similarities between flow-magnitudes with respect to time are considered.

## 11.3 MODELING TECHNIQUES

We divide the modeling techniques as supervised and unsupervised. In supervised, the training data contain both normal and anomalous videos, while unsupervised training data contain only normal videos.

### 11.3.1 SUPERVISED MODELS

#### *11.3.1.1 Shallow models*

There were many works carried out based on shallow models with simple handcrafted features given as input to a classifier. One such work was done by Wang and Snoussi [9], in which a histogram of optical flow orientation was introduced as a descriptor that was then fed to a one-class SVM for classification. Further, Zhang et al. [10] proposed an algorithm that used motion-improved Weber local descriptor (MoIWLD) for capturing low-level features and then gave it to a sparse-representation-based classifier. The proposed approach showed superior performance on three benchmark data sets for violence detection.

##### 11.3.1.1.1 Support vector machine

All the data points that are nearest to the hyperplane, which on altering changes the position of the dividing hyperplane, are called *support vector.* A hyperplane is a plane of dimension one less than the dimension of data space, which divides the classes of data. SVM is a learning algorithm mainly used on classification problems, which considers the data as support vectors and generates a hyperplane to classify them. There are three major kernels used in an SVM: linear, polynomial, radial basis function (RBF). The linear kernel is useful when the data are linearly separable, whereas the polynomial kernel is more suitable for data that can be separated by a curve of polynomial degree. The RBF kernel is the one that uses the squared Euclidean distance between two vectors to generate the hyperplane. Hassner et al. [8] have represented the change in flow-vector magnitudes using the ViF descriptor and detected violence using a linear SVM.

#### *11.3.1.2 Deep models*

Recently, the approach of deep learning models in computer vision and anomaly detection has been of great significance. In a work done by Ionescu et al. [11], they have differentiated two consecutive video sequences by using a binary classifier, which is trained iteratively. At each step, the classifier removes the most discriminant features, thereby helping the classifier to discriminate them more effectively. Another work done by Tran and Hogg [12] uses a convolutional autoencoder

(CAE), which extracts motion-feature, encodes it, and provides as input to a one-class SVM. To obtain a sparsity of higher degree, a winner-take-all step is brought in after the encoding layer. Further inspired by the strong feature learning ability of the convolutional neural networks (CNNs), Smeureanu et al. [13] extracted deep learning features using a pretrained CNN and an SVM for classification.

###### 11.3.1.2.1 Artificial neural networks

ANNs [14] or simply neural networks are one of the main methods used for classification and are inspired from the working of the human brain. An ANN has one input layer, one output layer, and one or more hidden layers. More and more hidden layers are used for learning more complex features. The architecture of an ANN is shown in Fig. 11.2. Each node in a layer has a vector of weights and an activation function through which data are transmitted for further learning. There are two main phases in the learning process of a neural network: forward and backward propagation. When the training data are fed into the network, it calculates the predicted output and compares it with the true output. An error is generated in the output layer based on this comparison, which is transmitted to the previous layer. With respect to this error received by each layer, the weights of each node are tuned.

###### 11.3.1.2.2 Convolutional neural networks

Similar to a neural network, CNN [15] also receives inputs through layers and has nodes through which this information is passed through. But its layers are more specialized and can accept volumes of data (image and video) unlike a simple neural network. This network comprises of four types of layers: convolution, ReLu, pooling, and fully connected (FC). In the convolution layer, a filter or a kernel is slid over the volume and convolution operation is applied to obtain an activation

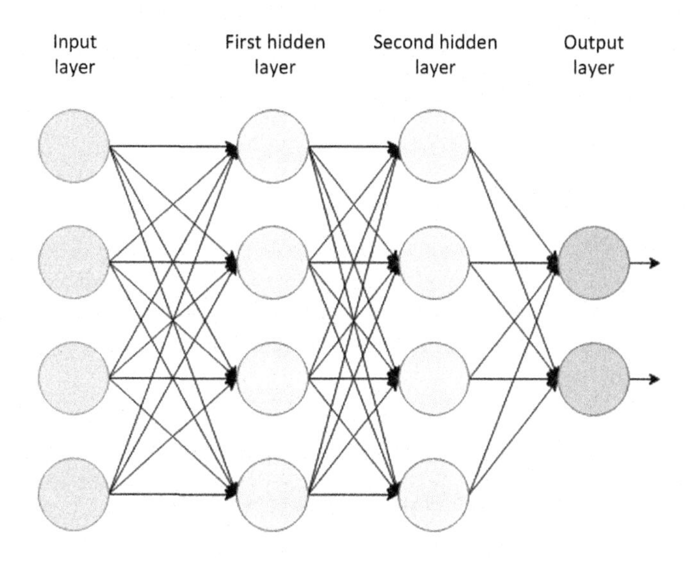


**FIGURE 11.2**

A simple ANN architecture.

map. Then this map is passed to the ReLu layer to increase nonlinearity. The resultant volume is fed to the pooling layer, which is used to capture the important features from the previous layers. The pooling layer can be either a maxed-pooling or an averaged pooling layer. The matrix is then flattened to obtain a one-dimensional column vector, which is then fed to the FC layer. The FC layer is used for classification purpose. One of the most significant CNNs is the AlexNet whose architecture is given in Fig. 11.3.

#### 11.3.1.2.3 Long short-term memory

Long short-term memory (LSTM) [16] networks are a special kind of recurrent neural networks that are capable of selectively remembering patterns for long duration of time. It is an ideal choice to model sequential data and hence used to learn complex dynamics of human activity. The long-term memory is called the cell state. Due to the recursive nature of the cells, previous information is stored within it. The forget gate placed below the cell state is used to modify the cell states. The forget gate outputs values saying which information to forget by multiplying 0 to a position in the matrix. If the output of the forget gate is 1, the information is kept in the cell. The input gates determine which information should enter the cell states. Finally, the output gate tells which information should be passed on to the next hidden state.

Two of the important variations for the LSTM model are deep LSTM (DLSTM) and CLSTM. DLSTM differs from the general LSTM in the number of layers the model contains. A single-layer LSTM will not be able to obtain well-defined temporal information. However, when more layers are stacked in the LSTM model, it will be able to acquire better temporal features, and hence will be more suitable in capturing motion in the time dimension [16]. In CLSTM, the data are first passed through convolutional layers, which ensure in capturing the spatial features, as shown in Fig. 11.4. The output from the CNN is provided to the LSTM, which will get the temporal features, and hence, the model will capture a motion with respect to both space and time [4]. These two variants can also be combined to give a convolutional deep LSTM, where the outputs from a CNN are given to a multilayer stacked LSTM [16], which is guaranteed to provide a better result at the cost of increased computational complexity.

## 11.3.2 UNSUPERVISED MODELS

### 11.3.2.1 Shallow models

A work done by Xiao et al. [18] employed a spatio-temporal pyramid, which captured the spatial and temporal continuities and also used a local coordinate factorization to tell whether a video is anomalous. Cheng et al. [19] presented a method with hierarchical feature representation to detect

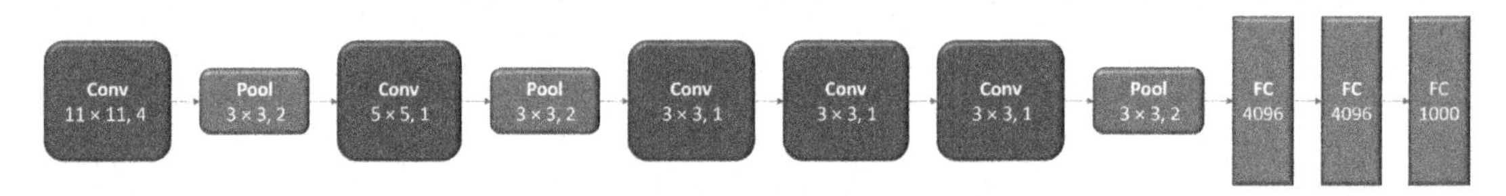


**FIGURE 11.3**

Illustration of the AlexNet architecture used for image recognition [15].

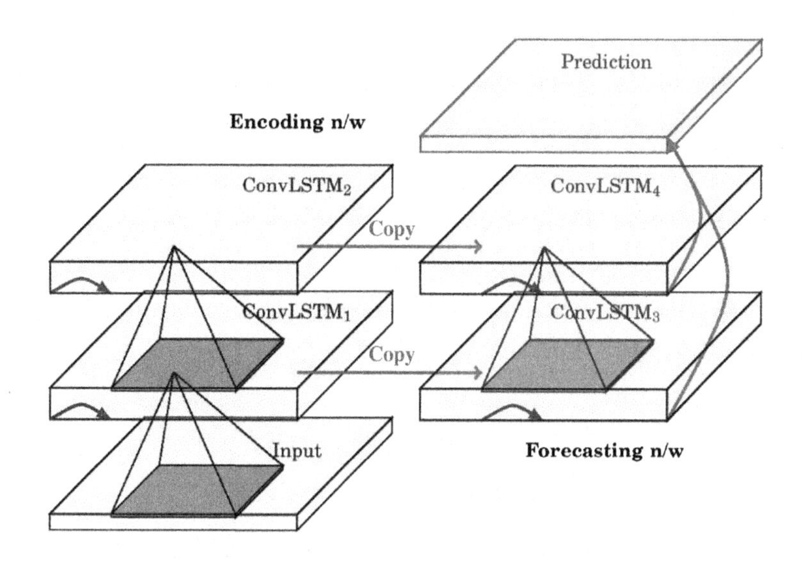


**FIGURE 11.4**

Structure of a convolutional long short-term memory [17].

local and global anomalies simultaneously by finding the relations between nearby sparse spatio-temporal interest points that were modeled by a Gaussian process regression. A popular work by Bermejo et al. [20] introduced the Hockey Fight data set, where violence detection was performed by taking spatio-temporal interest points (STIP) and motion scale invariant feature transform (MoSIFT) as action descriptors and bag-of-words (BoW) for fight detection. Besides, Leyva et al. [21] used the Gaussian mixture model (GMM), Markov chains, and BoW to prepare a compact feature set through which abnormal events are detected.

### 11.3.2.1.1 Principal component analysis

Principle component analysis is a form of representation learning model, which reconstructs the data to a lower dimension from the given training data and learns by reducing the reconstructed error. This is also used for dimensionality reductions and feature extractions, since principal component analysis (PCA) is known for its ability to extract the important features and still maintain the integrity of the original data. It computes eigenvectors by finding the covariance matrix of the standardized data. These vectors provide the variance directions and help in reconstruction.

For videos, PCA is used for modeling the spatial correlations between each pixel of a frame from its corresponding vector. In the case of anomaly or violence detection, the vector obtained will be of lower dimension and this captures the anomalous behavior. As each frame is associated with an optical flow value, this can be used for evaluating the reconstruction error. Kim and Grauman [22] used a probabilistic principal component analyzer, which captured the typical optical flow and, thereby, also learning normal patterns. The complex and costly step in this model is the optical flow estimation.

#### *11.3.2.2 Deep models*

Huang et al. [23] extracted low-level features including visual, motion-map, and energy features. Also mid-level features were extracted using a restricted Boltzmann machine (RBM) and deep representations of the crowd patterns were learned for the detection of unusual events. Recently, Sultani et al. [24] have introduced a new large data set comprising of surveillance videos. They have performed anomaly detection on this data set by segregating the normal and abnormal videos into bags and considering video sequences as instances for multiple instance learning. Another recent paper includes a method proposed by Ravanbakhsh et al. [25] uses generative adversarial nets (GANs) trained on normal frames and then used for abnormality detection. Likewise, Vu et al. [26] proposed a method where data representation was learned using an RBM followed by the reconstruction of the data. Based on the reconstruction errors, abnormal events were detected. In addition to the above methods, considering the importance of violence detection in video surveillance, Zhou et al. [27] trained FightNet by using image acceleration field as their input modal, which helps in capturing better motion features.

##### 11.3.2.2.1 Generative adversarial network

GAN is a model that is generative in nature as it uses joint probability distribution. GAN comprises of a generator and a discriminator. A generator constructs a fake sample from the given noisy training data. This fake sample is fed along with the stream of other training samples to the discriminator. The discriminator is similar to that of a binary classifier and classifies the training data as real or fake by assigning a probability to it. The generator is said to train on mapping the training data distribution and the discriminator trains on maximizing probability of assigning "real" label to the real training samples.

GANs can be easily used on videos for anomaly and violence detection by using the frames as training data. They evaluate a probability density distribution on the training set, which contains no anomalies, and provide an anomaly score that is the probability whether the sample is from the generator and thereby classifying it as an anomaly. GANs achieve this implicitly by minimizing the distance between the generative model and the training data distribution without the use of a parametric loss function. The mapping in the generator is done by transforming the image domain of the frames to a latent representation space. The loss from the discriminator is used in the backpropagation process of both the generator to generate images similar to the training samples and in the discriminator to classify the samples better. Ravanbakhsh et al. [28] used a modified version of GAN to produce the state-of-the-art results. They proposed a cross-channel GAN, as shown in Fig. 11.5, where the generator network is split into two parts: one to generate optical flow from frames and another to generate frames from the optical flow. The discriminator trains on both the generations, and hence, their method modeled a spatio-temporal correlation among the channels for better predictions.

##### 11.3.2.2.2 Autoencoders

Autoencoders are alternatives to PCA used for the purpose of dimensionality reduction by decreasing the reconstruction error on the training data. It is a neural network, which is trained by backward propagation. It performs a linear pointwise transform of the input using transformation

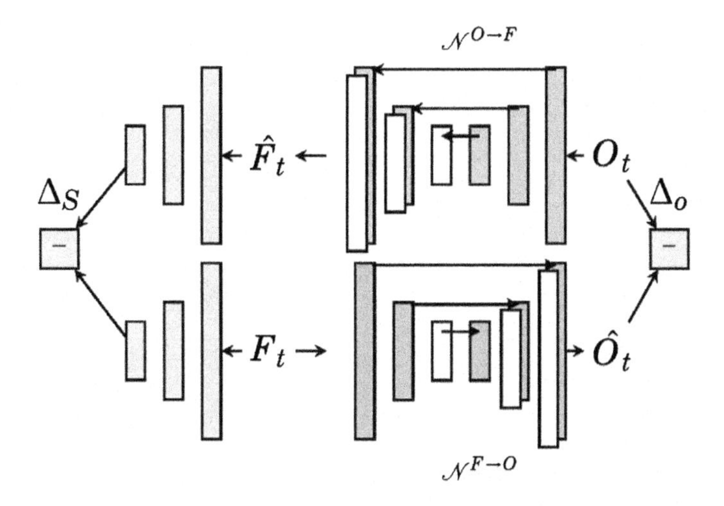


**FIGURE 11.5**

Cross-channel mechanism used in generative adversarial net for abnormality detection [17].

functions like ReLU or Sigmoid. Two of the main types of autoencoders are CAEs and 3D autoencoder.

***Convolutional autoencoder.*** The input signal is viewed as a signal that is decomposed as the sum of other signals by a normal autoencoder. This decomposition is made explicit by CAEs. CAEs are a type of CNNs. However, instead of manually assigning filters, we let the model learn optimal filters that minimize the reconstruction error. These filters can then be used to extract features from any input. Therefore CAEs are general-purpose feature extractors, which are trained only to learn filters capable of extracting features that can be used to reconstruct the input.

In a work done by Hasan et al. [29], an input sequence of frames from a trained video set was reconstructed by using a deep CAE. This is otherwise called as a spatio-temporal stacked frame autoencoder (STSAE). The STSAE stacks the frame sequence with each frame treated as a different channel in the input layer to a CAE. The architecture of the CAE and that of a stacked autoencoder is depicted in Fig. 11.6 and Fig. 11.7).

***3D Autoencoder.*** As discussed in [30], while 2D ConvNets are appropriate for image recognition and detection tasks, they are incapable of capturing the temporal features of consecutive frames for video analysis tasks. For this purpose, 3D convolutional architectures, depicted in Fig. 11.8, are used in the form of autoencoders. The 3D convolutional feature maps are encoded by the 3D autoencoder to obtain representations, which are invariant to spatio-temporal changes.

## 11.4 EXPERIMENTAL STUDY AND RESULT ANALYSIS

In this chapter, having security as a concern, we have studied extensively on violence recognition as it is regarded to be the most important section in anomaly detection. Although appearance features are prominently used, motion features have proven to be more effective in violence detection

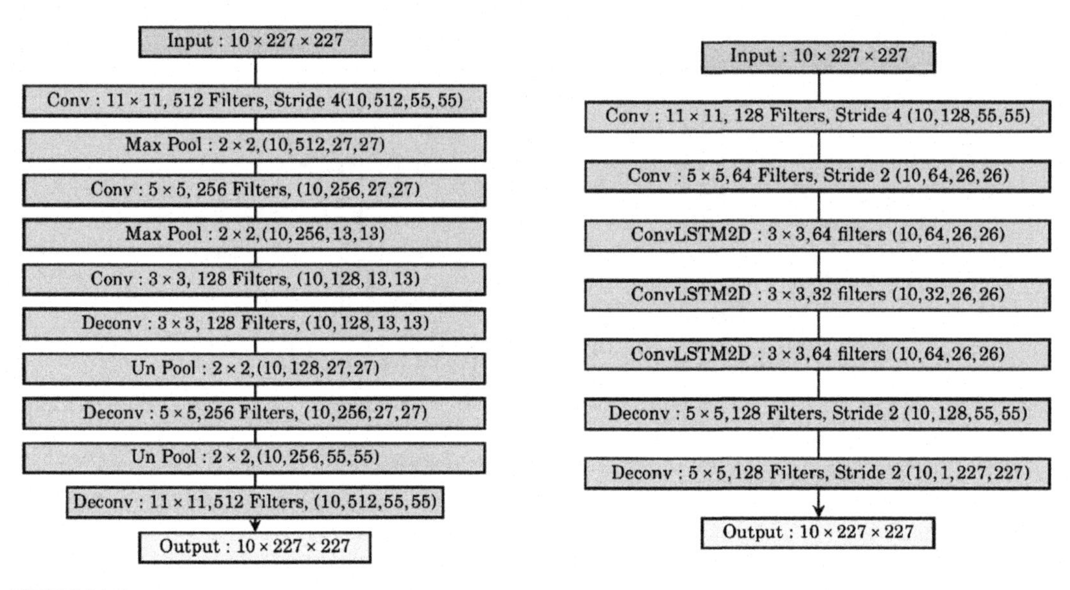


**FIGURE 11.6**

A stacked spatio-temporal autoencoder (left) and a convolutional long short-term memory autoencoder (right) for abnormal event detection [17].

task as appearance features sometimes might degrade the performance of the classifier. So in this study, we have focused on HOF feature along with SVM and ANN as classifiers for our experimentation.

## 11.4.1 DATA SETS

Our study is conducted on two standard benchmark challenging data sets: Hockey Fight and Crowd Violence data sets. The Hockey fight data set comprises of a total of 1000 video clips categorized as fight and no fight from the National Hockey Leagues. Each category consists of 500 video clips and thereby having 500 violent and 500 nonviolent clips. Each clip exactly consists of 50 frames having resolution of $360 \times 288$ pixels for each frame.

The Crowd Violence data set is specialized to test violence detection based on a crowd behavior. These videos characterize the violent and nonviolent behavior of crowd in public places, making it suitable for surveillance task. Crowd Violence has a total of 246 real video clips, of which 123 are violent and 123 are nonviolent with each frame having a resolution of $320 \times 240$ pixels.

## 11.4.2 COMPARATIVE STUDY ON RELATED WORK

The results of various recent state-of-the-art methods along with our basic study applied over Crowd violence and Hockey Fight data sets are shown in Tables 11.1 and 11.2. Shallow modeling techniques have proven to be effective in Crowd Violence data set. Due to the sparsely represented MoIWLD approach by Zhang et al. [10], there is minimal reconstruction and classification error,

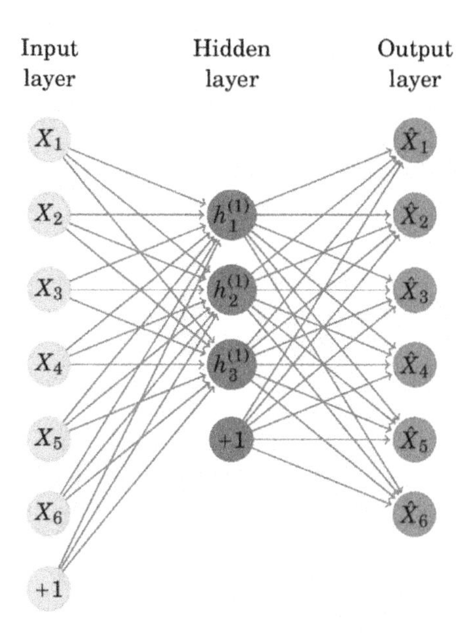


**FIGURE 11.7**

A simple autoencoder [17].

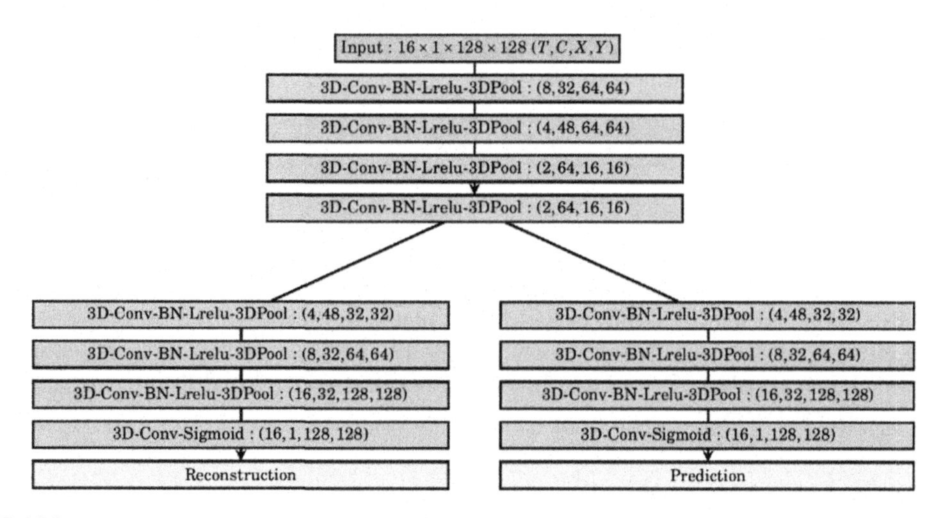


**FIGURE 11.8**

3D autoencoder architecture for video anomaly detection [17].

which provides a significant result in Crowd Violence. By capturing local and spatio-temporal features through IFV, Bilinski and Bremond [2] produced a state-of-the-art result in the Hockey Fight data set.

**Table 11.1 Results on Crowd Violence Data Set**

| Model | Method | ACC (%) |
|---|---|---|
| Shallow | IWLD [10] | 88.16 |
| | VIF + SVM [8] | 82.13 |
| | **HOF + SVM** | **83.37** |
| | **HOF + ANN** | **79.32** |
| Deep | GoogleNet + DLSTM [16] | 93.59 |
| | **HOF + ANN (dp = 0.1)** | **78.47** |
| | **HOF + ANN (dp = 0.13)** | **78.92** |

ANN, *Artificial neural network;* DLSTM, *deep long short-term memory;* dp, *dropout level;* HOF, *histogram of oriented optical flow;* IWLD; *improved Weber local descriptor;* SVM, *support vector machine;* ViF, *Violent Flows.*

**Table 11.2 Results on Hockey Fight Data Set**

| Model | Method | ACC (%) |
|---|---|---|
| Shallow | STIFV [2] | 93.40 |
| | STIP-HOG + HIK [20] | 91.70 |
| | **HOF + SVM** | **87.40** |
| | **HOF + ANN** | **87.13** |
| Deep | CLSTM [4] | 97.10 |
| | **HOF + ANN (dp = 0.1)** | **87.25** |
| | **HOF + ANN (dp = 0.13)** | **87.75** |

ANN, *artificial neural network;* CLSTM, *convolutional long short-term memory;* dp, *dropout level;* HOF, *histogram of oriented optical flow;* HOG, *histogram of oriented gradients;* STIP, *space–time interest point;* SVM, *support vector machine.*

We studied a baseline approach by using the HOF feature descriptor on both SVM and ANN models, where the results are obtained as shown in Table 11.1. Deep features even though being less explored have proven to show promising results on these data sets. Through the combination of GoogleNet Inception V3 CNN and stacked LSTM methods, Zhuang et al. [16] outperforms certain state-of-the-art results for Crowd Violence data set. Sudhakaran and Lanz [4] proposed a CLSTM which was capable of extracting low-level localized features and thereby reducing false alarm rate to a great extent.

### 11.4.3 OUR BASELINE STUDY

Some of the handcrafted feature descriptors and the classifiers mentioned in Section 11.3 are used for conducting this study. The accuracies attained for these models, for each data set, are shown in Tables 11.1 and 11.2. The results of our experiments are shown in bold. For both the models, features extracted remain the same, that is, the features extracted from HOF are given to a classifier. For modeling, we have considered SVM and ANN with one hidden layer as shallow, while the ANN with two hidden layers as deep. As mentioned earlier, if handcrafted features are given to a shallow network classifier, it represents a shallow model, while when provided to a deep network, it can be said to represent a deep model. We have conducted our study solely based on the above statement. HOF features are used to obtain the optical flow information. These features extracted are provided to an SVM and also to an ANN, whose parameters are determined by hyperparameter tuning, for classification based on shallow model representation. The features extracted for shallow models above are also used for studying deep models by providing them to an ANN consisting of two hidden layers, which is treated as a deep network. This network was trained on two different dropout levels to study the impact of it on the model.

We conducted experiments by using the method of $k$-fold cross-validation, where $k$ is five, that is, each data set is divided into five divisions each containing both the violent and nonviolent video clips. Features are extracted for each fold separately from each feature descriptor.

Training is performed by considering 80% as training set and the other 20% as test set. The average of all the accuracy in the fivefold validation is said to be the accuracy of the model.

Confusion Matrix for the second fold:

$$\text{Crowd Violence:} \quad \underset{\text{(Shallow)}}{\begin{bmatrix} 23 & 2 \\ 4 & 21 \end{bmatrix}} \quad \underset{\text{(Deep)}}{\begin{bmatrix} 22 & 3 \\ 7 & 18 \end{bmatrix}}$$

$$\text{Hockey Fight:} \quad \underset{\text{(Shallow)}}{\begin{bmatrix} 90 & 10 \\ 8 & 92 \end{bmatrix}} \quad \underset{\text{(Deep)}}{\begin{bmatrix} 97 & 3 \\ 6 & 94 \end{bmatrix}}$$

From the above obtained confusion matrix from one of the folds, it could be inferred that the Crowd Violence data set works better with shallow models as the false alarm rate is bound to be higher in deep models. This might be due to the fact that Crowd Violence being a small data set does not work well with deep networks. In contrast to the above, the Hockey Fight data set proves to work well with deep networks, since it has lesser false alarm rate. This is because of the large volume of data available in this data set compared with Crowd Violence. Our experimental study on deep networks was done by providing HOF features to a deep ANN model with two variations in dropout level. It can be seen that this baseline study with an HOF descriptor is more effective with Hockey Fight than that with Crowd Violence on both the methods. On further tuning, ANN may produce better results than other shallow methods.

## 11.5 CONCLUSION

Violence detection is one of the most important and essential tasks of video surveillance. In this chapter, we have focused on shallow and deep modeling techniques on two standard benchmark data sets such as Crowd Violence and Hockey Fight. We have done a comparative study of shallow and deep models on the above data sets and also on other state-of-the-art approaches for different feature descriptors and analyzed their results. In this chapter, we have inferred that for a small data set, shallow models perform well, but for a large data set, deep models give comparatively better performance at the cost of training time complexity.

## REFERENCES

[1] K. Lloyd, P. Rosin, D. Marshall, S. Moore, Detecting violent and abnormal crowd activity using temporal analysis of grey level co-occurrence matrix (GLCM)-based texture measures, Mach. Vis. Appl. (2017) 28, pp.361-371.

[2] P. Bilinski, F. Bremond, Human violence recognition and detection in surveillance videos, in: 2016 13th IEEE International Conference on Advanced Video and Signal Based Surveillance (AVSS), Colorado Springs, CO, 2016, pp. 30–36.

[3] Y.S. Chong, Y.H. Tay, Abnormal event detection in videos using spatiotemporal autoencoder, in: F. Cong, A. Leung, Q. Wei (Eds.), Advances in Neural Networks - ISNN 2017. ISNN 2017. Lecture Notes in Computer Science, vol. 10262, Springer, Cham, 2017, pp.189–196.

[4] S. Sudhakaran, O. Lanz, August. Learning to detect violent videos using convolutional long short-term memory. In *2017 14th IEEE International Conference on Advanced Video and Signal Based Surveillance (AVSS)* 2017 (pp. 1–6). IEEE.

[5] I. Laptev, T. Lindeberg, Space-time interest points, Int. J. Comput. Vis.- IJCV 64 (2003) 432–439. Available from: https://doi.org/10.1109/ICCV.2003.1238378.

[6] I. Laptev, On space-time interest points, Int. J. Comput. Vis. 64 (2005) 107–123. nos. 2–3.

[7] I. Laptev, M. Marszalek, C. Schmid, B. Rozenfeld, Learning realistic human actions from movies, CVPR (2008).

[8] T. Hassner, Y. Itcher, O. Kliper-Gross, Violent flows: real-time detection of violent crowd behavior, in: 2012 IEEE Computer Society Conference on Computer Vision and Pattern Recognition Workshops, Providence, RI, 2012, pp. 1–6.

[9] T. Wang, H. Snoussi, Detection of abnormal visual events via global optical flow orientation histogram, IEEE Trans. Inf. Forensics Security 9 (2014) 988–998.

[10] T. Zhang, W. Jia, X. He, J. Yang, Discriminative dictionary learning with motion weber local descriptor for violence detection, IEEE Trans. Circuits Syst. Video Technol. 27 (3) (2017) 696–709.

[11] R.T. Ionescu, S. Smeureanu, B. Alexe, M. Popescu, Unmasking the abnormal events in video. In *Proceedings of the IEEE International Conference on Computer Vision* 2017. (pp. 2895–2903).

[12] H.T. Tran, D. Hogg, September. Anomaly detection using a convolutional winner-take-all autoencoder. In *Proceedings of the British Machine Vision Conference 2017*. British Machine Vision Association, 2017.

[13] S. Smeureanu, R.T. Ionescu, M. Popescu, B. Alexe, Deep appearance features for abnormal behavior detection in video, in: S. Battiato, G. Gallo, R. Schettini, F. Stanco (Eds.), Image Analysis and Processing—ICIAP 2017. Lecture Notes in Computer Science, vol. 10485, Springer, Cham, 2017.

[14] M. Mishra, M. Srivastava, A view of artificial neural network, in: 2014 International Conference on Advances in Engineering & Technology Research (ICAETR - 2014), Unnao, 2014, pp. 1–3.

[15] H. Shin, et al., Deep convolutional neural networks for computer-aided detection: CNN architectures, dataset characteristics and transfer learning, IEEE Trans. Med. Imag. 35, 2016, pp. 1285–1298.

[16] N. Zhuang, J. Ye, K.A. Hua, Convolutional DLSTM for crowd scene understanding, in: 2017 IEEE International Symposium on Multimedia (ISM), Taichung, 2017, pp. 61–68.

[17] B.R. Kiran, D.M. Thomas, R. Parakkal, An overview of deep learning based methods for unsupervised and semi-supervised anomaly detection in videos, J. Imaging 4 (2018) 36.

[18] T. Xiao, C. Zhang, H. Zha, F. Wei, Anomaly detection via local coordinate factorization and spatio-temporal pyramid, in: D. Cremers, I. Reid, H. Saito, M.H. Yang (Eds.), Computer Vision—ACCV 2014. ACCV 2014. Lecture Notes in Computer Science, vol. 9007, Springer, Cham, 2015, pp. 66–82.

[19] K. Cheng, Y. Chen, W. Fang, Video anomaly detection and localization using hierarchical feature representation and Gaussian process regression, in: 2015 IEEE Conference on Computer Vision and Pattern Recognition (CVPR), Boston, MA, 2015, pp. 2909–2917.

[20] E. Bermejo Nievas, O. Deniz Suarez, G. Bueno García, R. Sukthankar, Violence detection in video using computer vision techniques, in: P. Real, D. Diaz-Pernil, H. Molina-Abril, A. Berciano, W. Kropatsch (Eds.), Computer Analysis of Images and Patterns. CAIP 2011. Lecture Notes in Computer Science, vol. 6855, Springer, Berlin, Heidelberg, 2011, pp. 332–339.

[21] R. Leyva, V. Sanchez, C. Li, Video anomaly detection with compact feature sets for online performance, IEEE Trans. Image Process. 26 (2017) 3463–3478.

[22] J. Kim, K. Grauman, Observe locally, infer globally: a space-time mrf for detecting abnormal activities with incremental updates, in: Computer Vision and Pattern Recognition, 2009. CVPR 2009. IEEE Conference on IEEE, 2009, pp. 2921–2928.

[23] S. Huang, D. Huang, X. Zhou, Learning multimodal deep representations for crowd anomaly event detection, Math. Probl. Eng. 2018 (2018) 13.

[24] W. Sultani, C. Chen, M. Shah, Real-world anomaly detection in surveillance videos, arXiv:1801.04264 [cs.CV], 2018.

[25] M. Ravanbakhsh, M. Nabi, E. Sangineto, L. Marcenaro, C. Regazzoni, N. Sebe, Abnormal event detection in videos using generative adversarial nets, in: 2017 IEEE International Conference on Image Processing (ICIP), Beijing, 2017, pp. 1577–1581.

[26] H. Vu, T.D. Nguyen, A. Travers, S. Venkatesh, D. Phung, Energy-based localized anomaly detection in video surveillance, in: J. Kim, K. Shim, L. Cao, J.G. Lee, X. Lin, Y.S. Moon (Eds.), Advances in Knowledge Discovery and Data Mining. PAKDD 2017. Lecture Notes in Computer Science, vol. 10234, Springer, Cham, 2017, pp. 641–653.

[27] P. Zhou, Q. Ding, H. Luo, X. Hou, Violence detection in surveillance video using low-level features, PLoS one 13 (10) (2018) e0203668.

[28] M. Ravanbakhsh, E. Sangineto, M. Nabi, N. Sebe, Training adversarial discriminators for cross-channel abnormal event detection in crowds, CoRR, vol. abs/1706.07680, 2017.

[29] M. Hasan, J. Choi, J. Neumann, A.K. Roy-Chowdhury, L.S. Davis, Learning temporal regularity in video sequences, in: Proceedings of the IEEE Conference on Computer Vision and Pattern Recognition, 2016, pp. 733–742.

[30] Y. Zhao, B. Deng, C. Shen, Y. Liu, H. Lu, X.-S. Hua, Spatio-temporal autoencoder for video anomaly detection, Proceedings of the 2017 ACM on Multimedia Conference Series MM'17, ACM, New York, 2017, pp. 1933–1941.

CHAPTER

# FPGA-BASED DETECTION AND TRACKING SYSTEM FOR SURVEILLANCE CAMERA

12

**Anitha Mary[1], Lina Rose[1] and Aldrin Karunakaran[2]**

[1]*Department of Instrumentation Engineering, Karunya Institute of Technology and Sciences, Coimbatore, India*

[2]*Department of Process Engineering, International Maritime College, Sultanate of Oman*

**CHAPTER OUTLINE**



## 12.1 INTRODUCTION

Object detection and tracking are important in surveillance, as it is required to observe the activity and report the information when there is a significant observed activity. Nowadays, surveillance systems are available commercially because of its robustness, decrease in hardware cost, and increase in processor speeds. Object detection using single camera is affected by many factors like occlusions, shadows, etc., and these shortcomings are handled by using multiple cameras. The modern machine vision algorithms can be easily processed using reconfigurable devices. The object can be detected using three methods, namely, temporal difference, optical flow, and background subtraction. Optical flow method has a disadvantage in meeting the requirement of real-time video processing. The temporal difference involves the difference in adjacent images depending on time sequence. The background difference, the video is captured using a static camera and this is converted into frames. The first frame is considered as reference frame and current frame as the frame under processing, and then it is given to subtraction operation. The output of operation is compared with the threshold $T$ to indicate the pixel movement from the object, as shown in Fig. 12.1. A good background subtraction has the capability to overcome varying illumination condition, shadows, etc.

Object tracking is another challenge task. It involves tracking of object of interest. The surveillance system is the process of monitoring the behavior, activities of people, and other changing

**The Cognitive Approach in Cloud Computing and Internet of Things Technologies for Surveillance Tracking Systems.**
**DOI: https://doi.org/10.1016/B978-0-12-816385-6.00012-X**

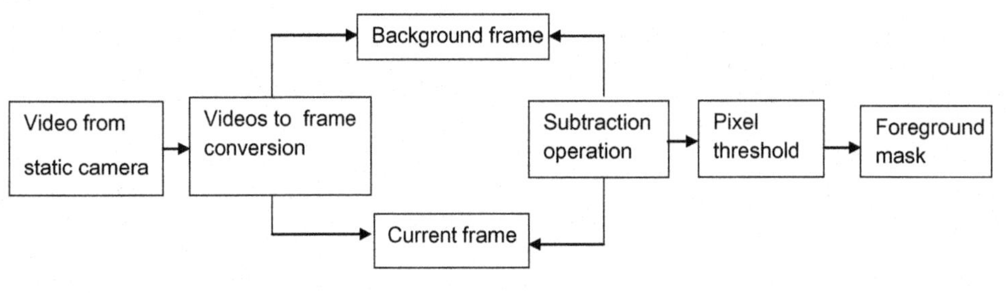


**FIGURE 12.1**

Background subtraction process.

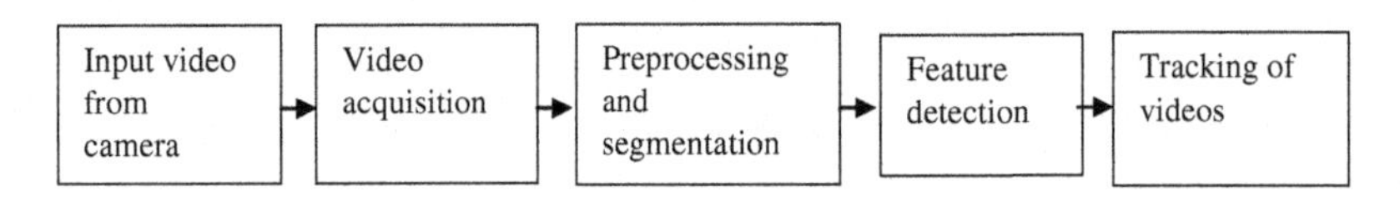


**FIGURE 12.2**

Video tracking process.

information. The system uses a static camera and there are three main processes, namely, object detection, tracking, and recognition. Fig. 12.2 shows the tracking process.

## 12.2 PRIOR RESEARCH

Video surveillance becoming very important in many areas as 24-h monitoring by humans is crucial. An intelligent transport system suggested by Buch et al. [1] is of low cost. Attard and Farrugia [2] proposed a surveillance system to detect humans and vehicles. For object detection, Huang [3] proposed moving object detection using the background model, which uses rapid matching algorithm to get optimum background scenes. Fradi and Dugelay [4] proposed uniform motion model for foreground segmentation. For video surveillance applications, the video data are initially compressed. This is achieved by minimizing the spatial and temporal frames, thereby improving the quality of the video. Motion detection is the one where moving objects are identified and motion estimation represents the position of the moving objects. Horn–Schunck algorithm uses optical flow vector calculation and is more sensitive to noise and found to be less efficient under occlusion conditions. Krattenthaler et al. [5] suggested preprocessing for color conversion. Wallhoff et al. [6] suggested reliable segmentation for surveillance recognition.

The better accuracy of the 3D imaging system is proved by Beder et al. [7]. Many image processing methods like feature-based, edge-based, and model-based object detection and tracking are given by Krattenthaler et al. [5], Stauffer and Grimson [8], Jain et al. [9], and Sheikh et al. [10] suggested that the background subtraction is best for extracting foreground image. Johnson and Tews [11] suggested effective classifier to determine the foreground images, which are influenced by background image. The background subtraction techniques give accurate results in stationary background environment. However, moving objects are obtained by using affine model but lacks in accuracy when the scene has significant depth variations. Implementation of computer vision algorithm in hardware is a time-consuming one. It requires high computing system called field programmable gate array (FPGA), which can handle different task in reconfigurable techniques.

## 12.3 SURVEILLANCE SYSTEM TASKS AND CHALLENGES

In general, the surveillance system should able to detect the presence of objects in the field of view, tracking these objects over time and classification based on activities and reporting information about the events happening within the field of view. The object detection and tracking poses various challenges with respect to system designers.

In the case of object detection, background subtraction is popular among other methods. Fig. 12.3 shows the example for background subtraction output, but it faces several problems in accurately detecting objects in realistic environments.

1. The change in illumination with daytime modifies the scene appearance, causing deviation from the background model. As a result, there will be increase in the falsely foreground images, which makes the system unrealistic.
2. It is not possible to differentiate the images when the current image is similar to the background image.
3. Shadow objects will be considered as foreground images, which results in false detection.

Once the object is detected, it is necessary to record the object information over time. When the objects are being tracked, its shape, size, or motion undergo various changes in realistic environment. Second, the objects undergoes occlusion, which means the object is being blocked by other objects or structure. When the object undergoes occlusion, it is difficult to find the position and velocity of an object. There are two types of occlusions as follows.

1. Interobject occlusion: blocking of one object with other. One of the examples is people moving in groups. There will be frequent interobject occlusion. Detecting such case is important for surveillance applications.
2. Scene structure occlusion: here, object disappears after a certain amount of time, for example, person walking behind the building.

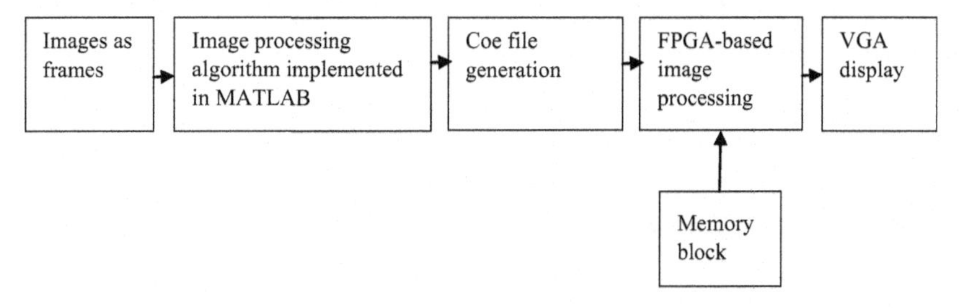


**FIGURE 12.3**

Proposed method for object detection in FPGA.

## 12.4 METHODOLOGY

The techniques used to track the target of interest are known as the video tracking of objects. It gains interest in security and surveillance system, medical imaging, and robotics. FPGAs are efficient, high-speed, and low-cost devices, which can process the image processing algorithms effectively. The objective of this chapter is to design the object tracking using FPGA. The static images are initially stored in FPGA memory. The image processing algorithm is implemented using VHSIC hardware description language (VHDL) and it is displayed using VGA display. The block diagram of the proposed system is shown in Fig. 12.4:

The input image of size $256 \times 256$ each of 8-bit wide is processed in MATLAB environment. The coe (coefficient) is generated and it is stored in the memory block of the FPGA. It is further processed using the FPGA and the resulting image is displayed in VGA display.

For object detection, the first step is to convert an image into grayscale conversion.

The image that is in RGB format is converted into grayscale, whose values are in the range of 0–255. The pseudocode that converts an image to grayscale image is given as follows:

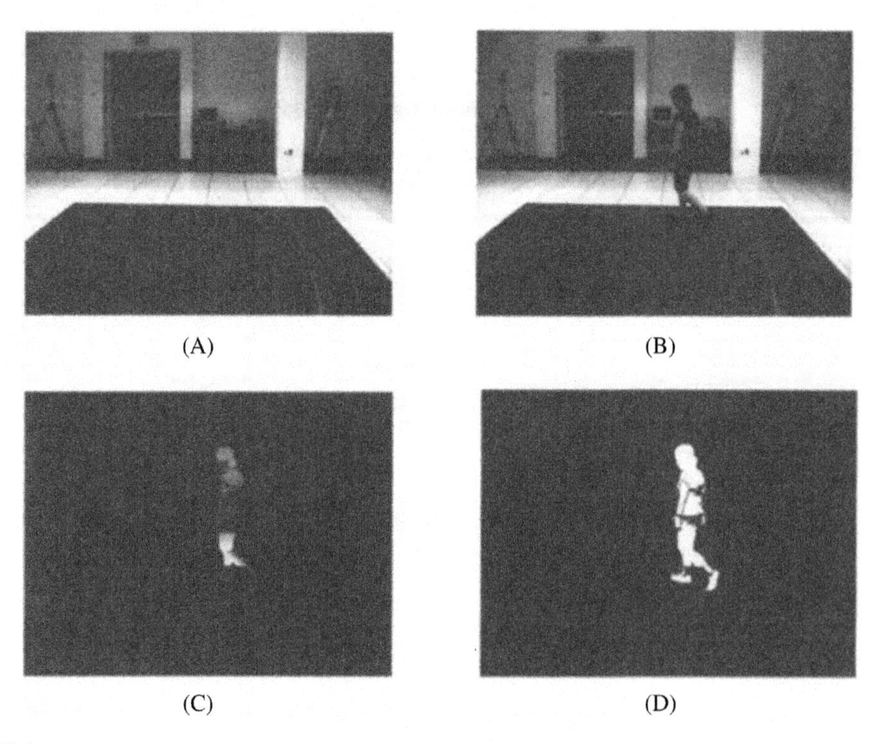


**FIGURE 12.4**

(A) Background image. (B) Foreground image. (C) Image subtraction. (D) Binary image.

```
Image1= imread ('C:\users\user1.jpg');
Image2=rgb2gray(Image1);
Y1=step(hautothresh, Image2);
F=double(Image);
```

After the grayscale conversion, the background frames are subtracted from the foreground image. The resulting foreground image is also called delta frame, which consists of less number of pixels and is easy to process. In order to increase the resolution of the image, thresholding is done.

The grayscale image consists of a number of pixels, and each pixel is called object pixels. If the object pixel is greater than the set threshold value, then the object pixel gains a value of "1" and the background image "0." In order to reduce the noise in the image, median filters are normally used. The filtered image is further subjected to edge detection using Sobel operators. Fig. 12.4 shows the binary subtraction operations.

The object detection gives the information about its shape and size. Now the frames are arranged in accordance with the video. One frame is processed in MATLAB per second. The center of the mass method is used to track the target. The image processing algorithm is implemented using MATLAB environment.

The pseudocode for .coe file generation in MATLAB is given as follows.

```
for row=1:height
For column=1:width
R=BW(row,column,1);
R=BW(row,column,1);
R=BW(row,column,1);
Y=[R;G;B];
```

Later, it is implemented using the Spartan 3E FPGA kit. Both spatial and temporal parallelisms can be implemented easily using FPGA. The images are first downsampled from an 8-bit RGB value to an 8-bit grayscale value. This is given in Eqs. (12.1) and (12.2)

$$\text{Grayscale} = (R \times 0.25) + (G \times 0.5) + (B \times 0.125) \tag{12.1}$$

The VHDL code for object detection using the distance equation

$$D = \text{sqrt}((\text{RI} - \text{RT})\hat{}2 + (\text{GI} - \text{GT})\hat{}2 + (\text{BI} - \text{BT})\hat{}2) \tag{12.2}$$

is given in the following

```
process(RI, BI, GI)
begin
 if (RI > RT) then
 Rdif := RI - RT;
 else
 Rdif := RT - RI;
 end if;
 if (GI > GT) then
 Gdif := GI - GT;
 else
```

```
  Gdif := GT - GI;
  end if;
  if (BI > BT) then
  Bdif := BI - BT;
  else
  Bdif := BT - BI;
  end if;
R2 := Rdif * Rdif;
G2 := Gdif * Gdif;
B2 := Bdif * Bdif;
sqrtsum<= (R2) + (G2) + (B2);
if (sqrtsum> threshold) then
  0 <= '0';
else
  0 <= '1';
end if;
end process;
endBehavioral;
```

Fig. 12.5 shows the center of mass calculation and Fig. 12.6 shows the dilation operation for object tracking.

Table 12.1 shows the synthesis report of object detection and tracking system.

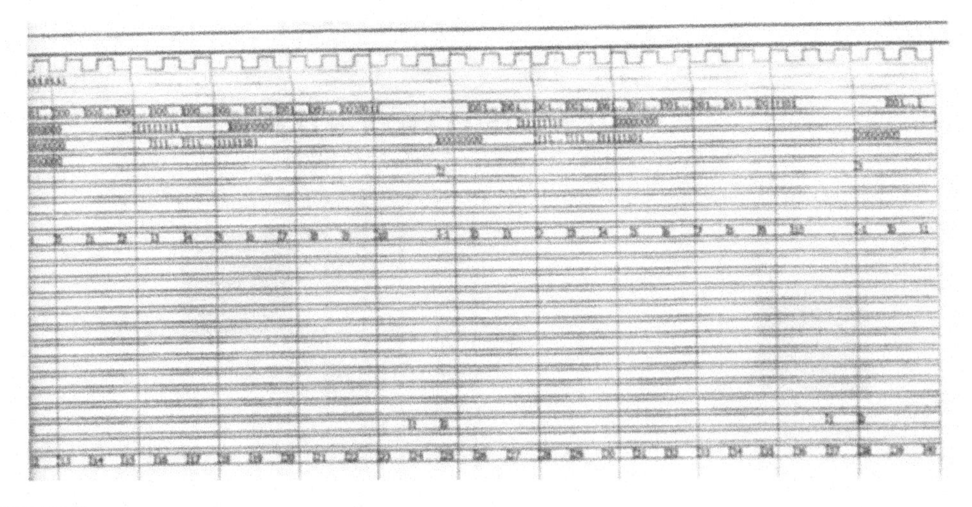

**FIGURE 12.5**

Center of mass calculation.

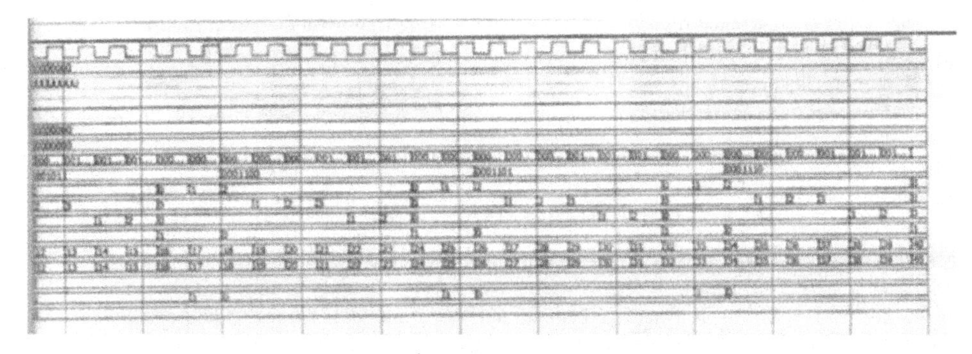

**FIGURE 12.6**

Dilation.

**Table 12.1 Synthesis Report.**

| Selected Devices | 3s250etq144-5 | |
|---|---|---|
| Number of slices | 2431 out of 2448 | 99% |
| Number of slice flip flops | 1763 out of 4896 | 36% |
| Number of input LUTs | 4192 out of 4896 | 85% |
| Number of IOs | 12 | |
| Number of bonded IOBs | 12 out of 108 | 11% |
| Number of BRAMs | 9 out of 12 | 75% |
| Number of GCLKs | 1 out of 24 | 4% |

*LUT,* Look up table; *IO,* Input output; *IOB,* Input output block; *BRAM,* Block random access memory; *GCLK,* Gobal clock.

## 12.5 CONCLUSION

Object detection and tracking find applications in autonomous robot navigation, and surveillance and vehicle navigation. Object detection involves obtaining objects in consecutive frames. Object tracking locates the objects to appear over time using a camera. An FPGA includes a number of configurable logic blocks, distributed memory, and hard digital signal processing modules, which has the capability of processing real-time objects. In this chapter, object identification and tracking are simulated using MATLAB environment and implanted using the SPARTAN 3E kit.

## REFERENCES

[1] N. Buch, S.A. Velastin, J. Orwell, A review of computer vision techniques for the analysis of urban traffic, IEEE Trans. Intell. Transport. Syst. 12 (3) (2011) 920–939.

[2] L. Attard, R.A. Farrugia, Vision based surveillance system [C], in: Proceedings of IEEE EUROCON-International Conference on Computer as a Tool (EUROCON 2011), Portugal 2011, pp. 1–4.

[3] S.C. Huang, An advanced motion detection algorithm with video quality analysis for video surveillance systems, IEEE Trans. Circuits Syst. Video Technol. 21 (1) (2011) 1–14.

[4] H. Fradi, J. Dugelay, Robust foreground segmentation using improved Gaussian mixture model and optical flow, in: Proceedings of IEEE International Conference on Informatics, Electronics & Vision (ICIEV-2012). Dhaka, Bangladesh, 2012, pp. 248–253.

[5] W. Krattenthaler, K.J. Mayer, M. Zeiler, Point correlation: a reduced const template matching technique, ICIP (1994) 208–212.

[6] F. Wallhoff, M. Ruß, G. Rigoll, J. Gobel, H. Diehl, Surveillance and activity recognition with depth information, in: IEEE International Conference on Multimedia and Expo, Beijing, China July 2007.

[7] C. Beder, B. Bartczak, R. Koch, A comparison of PMD-cameras and stereo-vision for the task of surface reconstruction using patchlets, in: IEEE Conference on Computer Vision and Pattern Recognition, Minneapolis, MN, 2007, pp. 1–8.

[8] C. Stauffer, W.E.L. Grimson, Adaptive background mixture models for real-time tracking, CVPR (1999) 252–260.

[9] A.K. Jain, Y. Zhong, S. Lakshmanan, Object matching using deformable templates, PAMI 18 (3) (1996) 267–278.

[10] Y. Sheikh, O. Javed, T. Kanade, Background subtraction for freely moving cameras, in: Proceedings of the IEEE 12th International Conference on Computer Vision, Kyoto, Japan, 2009, pp. 1219–1225.

[11] S. Johnson, A. Tews, Real-time object tracking and classification using a static camera, in: Proceedings of the IEEE International Conference on Robotics and Automation (ICRA '09), Kobe, Japan, 2009.

## FURTHER READING

A. Acasandrei, A. Barriga, Design methodology for face detection acceleration, in: IEEE Conference of the Industrial Electronics Society (IECON), 2013. Vienna, Austria.

U. Ali, M.B. Malik, K. Munawar, FPGA/soft-processor based real-time object tracking system, in: Proceedings of IEEE, Fifth Southern Programmable Logic Conference, 2009, pp. 33–37.

X. Anitha Mary, D. Domnic, K. Rajasekaran, Analysis of resource utilization for a floating point complex multiplication in FPGA, J. VLSI Des. Tools Technol., 2010.

S. Anjana, X. Anitha Mary, Noise cancellation using adaptive filters in FPGA, Res. J. Recent. Sci. 3 (2014) 4–8. SNIP 1.33.

G.J. García, C.A. Jara, J. Pomares, A. Alabdo, L.M. Poggi, F. Torres, A survey on FPGA-based sensor systems: towards intelligent and reconfigurable low-power sensors for computer vision, control and signal processing, Sensors 14 (4) (2014) 6247–6278.

S.K. Jose, X. Anitha Mary, X. Namitha mathew, ARM 7 based accident alert and vehicle tracking system, Int. J. Innov. Technol. Explor. Eng. 2 (4) (2013).

J. Prabhakar, X. Anitha Mary, FPGA based lane deviation system using system generator, Int. J. Adv. Res. Comput. Sci. Softw. Eng. 3 (2) (2013).

K.S. Raju, G. Baruah, M. Rajesham, P. Phukan, Computing displacement of moving object in a real time video using EDK, in: International Conference on Computing, Communications, Systems and Applications (ICCCSA), Hyderabad, March 30–31, 2012, pp. 76–79. ISBN:978-81-921580-8-2.

M. Thilagavathi, X. Anitha Mary, FPGA based quadrature decoder/counter with I2C bus interface for motor control, J. Comput. Sci. 4 (4) (2010).

M.J. Thilagavathi, X. Anitha Mary, Implementation of FPGA based incremental PID controller using conventional method and distributed arithmetic, J. Comput. Sci. 4 (5) (2010).

S. Zafeiriou, C. Zhang, Z. Zhang, A survey on face detection in the wild: past, present and future, J. Comput. Vis. Image Underst. (2015).

Y. Zhang, X. Tong, T. Yang, W. Ma, Multi-model estimation based moving object detection for aerial video, Sensors 15 (4) (2015) 8214–8231.

# Index

*Note*: Page numbers followed by "*f*" and "*t*" refer to figures and tables, respectively.

## D



## E



## F



## G



## H

## S

## W



## Z